PHILOSOPHY IN ECONOMICS

THE UNIVERSITY OF WESTERN ONTARIO SERIES IN PHILOSOPHY OF SCIENCE

A SERIES OF BOOKS
IN PHILOSOPHY OF SCIENCE, METHODOLOGY,
EPISTEMOLOGY, LOGIC, HISTORY OF SCIENCE,
AND RELATED FIELDS

VOLUME 16

PHILOSOPHY IN ECONOMICS

Papers Deriving from and Related to a Workshop on Testability and Explanation in Economics held at Virginia Polytechnic Institute and State University, 1979

Edited by

JOSEPH C. PITT

Department of Philosophy and Religion
Virginia Polytechnic Institute and State University
Blacksburg, Virginia

D. REIDEL PUBLISHING COMPANY

DORDRECHT : HOLLAND / BOSTON : U.S.A.

LONDON : ENGLAND

Library of Congress Cataloging in Publication Data

CIP

Workshop on Testability and Explanation in Economics, Virginia Polytechnic Institute and State University, 1979.
Philosophy in economics.

(The University of Western Ontario series in philosophy of science ; v. 16)
Includes bibliographies and index.
1. Economics—Congresses. 2. Economics—Philosophy—Congresses. 3. Economics—Methodology—Congresses. I. Pitt, Joseph C. II. Title. III. Series: University of Western Ontario. University of Western Ontario series in philosophy of science ; v. 16.

HB21.W67 1979 330'.01 80–27996
ISBN-13: 978-94-009-8396-0 e-ISBN-13: 978-94-009-8394-6
DOI: 10.1007/978-94-009-8394-6

Published by D. Reidel Publishing Company,
P.O. Box 17, 3300 AA Dordrecht, Holland

Sold and distributed in the U.S.A. and Canada
by Kluwer Boston Inc.,
190 Old Derby Street, Hingham, MA 02043, U.S.A.

In all other countries, sold and distributed
by Kluwer Academic Publishers Group
P.O. Box 322, 3300 AH Dordrecht, Holland

D. Reidel Publishing Company is a member of the Kluwer Group

PREFACE

The essays in this volume are the result of a workshop held at Virginia Polytechnic Institute and State University in April, 1979. The assembled group was diverse, comprised of philosophers, economists, and statisticians. But it was not the complete group on which we had initially planned. Richard Rudner was in France on sabbatical and was unable to fly back for the occasion. His untimely death the following summer saddened us all, for we lost not just a colleague but a friend. This book is dedicated to him out of the spirit of friendship and in appreciation for the groundbreaking work he did in the philosophy of the social sciences.

In addition to the participants, a number of people worked very hard to make our gathering possible. We are especially indebted to Dean Henry Bauer, Dean Ernie Stout and Dean John Ballweg of the College of Arts and Sciences at Virginia Tech for their good will and support, both moral and substantive. We would also like to thank Professor Guy Hammond, Head, Department of Philosophy and Religion, for his council and assistance. Our special thanks to Jeanne Keister and Betty Davis for their patience with unending typing and reservations, and finally to Barbara Kersey, always at hand, ever helpful. Without them nothing would have transpired.

The final versions of the papers would not have been possible without the enormous cooperation of the authors themselves. With good spirit and incredible speed they not only commented on each other's work, but responded cheerfully to my constant requests for instant productivity.

Blacksburg, Virginia
June, 1980

J.C.P.

TABLE OF CONTENTS

Dedicated to
Richard R. Rudner

JOSEPH C. PITT

INTRODUCTION

For much of the second quarter of this century, logical positivism was the dominant philosophy of science in the Anglo-American world. It always was under fire, but gradually those attacks mounted in the 1950s and 1960s began to take their toll and the possibility of alternative programs began to be investigated. A number of such alternatives were promulgated and in the wake of their success in attracting devotees positivism seemed to wither on the vine. But if the history of thought teaches us anything, it is that not only are entire research programs rarely eliminated without a trace, but also that such large scale programs leave legacies and that it is often some time before they can be discerned.

It appears that we are just now beginning to see the legacy of logical positivism. Further more, if the papers in this volume are to be considered any kind of evidence at all concerning the ebb and flow of philosophical ideas, that legacy may come as something of a pleasant surprise. For we now find in philosophical discussions of, of all places, the social sciences, and in particular of economics, an awareness of many of the positivists more cherished concepts. Of distinct note is the general problem of the relation between pure and applied theory and the adjacent problem of testing hypotheses. It is these issues that the papers in this volume basically address.

These are not exactly the same concerns of Carnap, Neurath, and Reichenbach, for their objective was the construction of a heirarchically arranged axiomatized structure in which all the sciences could be logically interpreted, if not reduced. But legacy is not to be equated with direct descendance. It is rather what we are bequeathed. In this case, it is a vision of the goal of scientific inquiry: a systematic and rigorous body of knowledge.

Furthermore, the legacy entails more than merely the statement of a goal. We are also left with a number of specific concerns which, if examined in the context of specific areas, can help guide us toward the point where we are in a position to determine if we have achieved the goal. As noted above, there is one such concern in particular which the essays here help illuminate, the relation between pure theory and its applications. And while most of these papers deal expressly with economic theory or the problems associated with its application, the main philosophical issue behind the scenes remains the

J. C. Pitt (ed.), Philosophy in Economics, 1–4.

positivist's demand that an empirical theory be testable. As a general topic it has a number of dimensions.

We find, for example, in Green's paper, not only the distinction between pure and applied theory, but an attempt to show what bearing each has on the other. Hausman picks up this theme by examining the explanatory dimension of certain applications of what some have called the basic economic theory, namely, equilibrium theory. The major thrust of his argument concerns the amount of information we can expect to extract from an abstract mathematical structure and how it is affected when supplemented by auxiliary theory. Moving to a more specific example of this problem, Wolfson examines the consequences of a new theory of consumer behavior for our understanding of increase in simplicity in structure and increase in empirical fruitfulness.

Rosenberg offers a critique of goals and objectives of economic theory with the objective of raising in bold type the question of how much we can expect economic theory to produce. Dacey follows this up by giving a specific example of some limitations by exploring the fascinating phenomenon of theory absorption, the process whereby a consumer internalizes the precepts of a theory, thereby altering the ability of the social scientist to predict his behavior. The crux of this argument is the effect of theory absorption on our understanding of rational choice.

Strasnick extends the examination of this problem by considering the consequences of certain features of the initial assumptions employed in formulating the principles of von Neumann-Morgenstern theory of games, an issue examined further by McClennen. The results of these two papers illustrate the effect of initial theoretical assumption on specific outputs of a theory.

Wilde continues the examination of that same set of worries, only by proceeding from "the other end." He addresses the problem of formulating adequate experiments using laboratory animals and moves us closer to understanding the difficulties inherent in turning highly abstract mathematical theories into experimentally testable ones. The problem of hypothesis testing in general is reexamined by I. J. Good, while Mayo confronts the enormous problem of unraveling the complexities of using statistical methodologies.

Over and above noting that these essays are concerned with the limits of economics in terms of its initial assumptions, its testability, its explanatory scope, and simplicity, what else can we say of them? What else need we say? The kinds of questions addressed and the varied backgrounds of the writers speak to the durability and deep-grained character of these worries. It is important to remember that these papers are the result of an *interdisciplinary* workshop on testability in economics. The consequence is fascinating. We find that a full discussion of the problem reads almost like the table of contents from either Rudner's *Philosophy of Social Sciences* or Hempel's *Aspects of Scientific Explanation*. And if the authors were all philosophers, that would be understandable. But they are not. They come from areas as diverse as statistics, business management, and economics, in addition to philosophy.

There are two ways to read these results: skeptically and not so skeptically. On the skeptic's side, it could be observed that this result is due to what we might call 'intellectual time-lag', for it may be argued that the reason why these issues are brought up by non-philosophers is that philosophers have dealt with them for some time, and surely by now we have made some impact on non-philosophers. Furthermore, the reason why these are hot issues in the social sciences and not in the hard sciences is that they have been exhausted in the hard sciences and philosophically minded thinkers in those areas are now concerned with the really tough issues which require detailed knowledge of the specific theories.

In response, I would point out first that the details of economic theory contained in the papers here are hardly superficial. Second, it is not clear that problems of explanation and testing have been exhausted at any point. Finally, it may be that one reason why these issues are not being pursued in the physical sciences with as much vigor as in the social sciences concerns expected degree of complexity. Social scientists come to their disciplinary studies knowing that sooner or later they have to deal with the ultimate variable, man. Investigators concerned with the hard sciences may have assumed at some early point that without man their problems will be minimized. It just may be that currently they are suffering from the failure of that set of expectations.

But if we turn to the not so skeptical view of why traditional philosophical problems keep cropping up in studies of the social sciences, we may just have an answer that avoids questions of specific difficulties of disciplines, etc. These philosophical issues arise because the answers to them provide us with a constitutive account of the cognitive significance of scientific inquiry. To find that the same philosophical problems occur in discussion of the social sciences as in other sciences is one way of reassuring ourselves that even where human social interaction is the subject matter, scientific results are still possible.

In effect, what we see here is a strong testimonial to the depth of understanding that the logical positivists, with all their faults, had of science. Their specific answers may have failed, but their questions remain with us; so much so that they almost characterize the nature of inquiry about science itself.

In the long run, we may all be forced to the conclusion that there is no prescription for how to do philosophy of science. It may be that we can only learn by example. In that case, we hope that the example offered here furthers, rather than retards, the discussion by enlarging the range of possibilities for fruitful analysis.

Virginia Polytechnic Institute and State University

EDWARD J. GREEN*

ON THE ROLE OF FUNDAMENTAL THEORY IN POSITIVE ECONOMICS

1. INTRODUCTION

There is a standard distinction in economics between positive theory, which concerns law-like propositions about individual or aggregate economic behavior, and normative or welfare theory, which concerns the evaluation of alternative policies or institutions. In addition, contributions to economic theory are often classified according to whether they are 'pure' or 'applied'. Although this classification is not a precise distinction, there is substantial agreement about which contributions are quite 'pure' and which are rather 'applied'. The present paper concerns the relation between positive theories which lie toward opposite ends of this spectrum. In order to avoid some misleading implications of the traditional way of speaking, I will employ the term 'fundamental theory' to refer to contributions which are usually called 'pure', and 'specific theory' to describe those which are usually called 'applied'. Although I will offer definitions of these terms, I do not intend to posit a rigid distinction, of which every contribution must fall unambiguously on either one side or the other. Another understanding which should be explicit at the outset is that, in discussing the role of fundamental theory, I will be systematically vague about the philosophical theory of science. My point will be that fundamental theory performs functions and satisfies criteria which would be necessary on any reasonable view of science, rather than that some particular view can explain its existence and justify its usefulness. In particular, when I use terms like 'empirically significant' and 'research program', I intend these terms to be taken in a general way rather than in the technical senses which they have acquired in specific philosophical theories.

Specific theories in positive economics are easy to recognize as scientific theories. In these specific theories, models of particular markets are formulated, and empirically significant consequences are derived. Questions like why these consequences are significant (e.g., that they are verifiable or refutable) may be matters of debate, but these are questions about science in general and not questions about the scientific status of economics.

The parallel between fundamental economic theories and theories in the

J. C. Pitt (ed.), Philosophy in Economics, 5–15.

physical sciences is not so clear. Fundamental theories typically look like arguments from first principles to propositions devoid of empirical content: for instance, that a competitive equilibrium exists. Even within the economics profession, many have profound doubts about the value of such exercises. Concerning the existence of equilibrium in particular, many economists would subscribe to the view that "questions of nonexistent equilibria are virtually irrelevant. Clearly something *is* happening in the real world. The object of science is to explain these events. Dwelling on the possible non-existence of these events is of questionable empirical value." [10, pp. 526–527]. Even the terminology of 'pure' versus 'applied' theory seems to carry the implication that the former may be an essentially speculative endeavor.

That this implication is false, and that fundamental and specific economic theories are really complementary, is the thesis of this paper. The argument has two parts. First, fundamental theory is defined, and an account is given of how, in neoclassical positive economics, fundamental theory enhances the acceptability of specific theory as a research program. This account should establish that fundamental theory cannot be isolated from specific theory as easily as is implied by the skeptical view quoted above. However, the account might be turned to charge that neoclassical specific theories are unacceptable because they depend on 'unscientific' fundamental theory, rather than to argue that fundamental theory is 'scientific' because it is of a piece with specific theory. The second part of the argument here will forestall this move by pointing out that reliance on fundamental theory is not a peculiar feature of neoclassical economics, and by suggesting that standard views of science indicate ways in which fundamental theories can be evaluated systematically. The second part of the argument will reinforce the first part by establishing that both fundamental theories and specific theories must be included in any viable research program in economics, whether neoclassical or other.

2. THE USEFULNESS OF FUNDAMENTAL THEORY TO NEOCLASSICAL POSITIVE ECONOMICS

While specific theory has already been characterized in the introduction of this paper as the formulation and study of models of particular markets or market phenomena, fundamental theory has been described only as not being specific. In fact, it is typical for fundamental theory to be described either in this way, or else by enumeration of some fundamental theories, or else according to some *ad hoc* criterion (e.g., that fundamental theory uses advanced mathematics). In view of the lack of a clear conception of what fundamental theory is, it is not surprising that there is uncertainty about how it is related to specific theory or whether it is related at all. A definition of fundamental theory will be offered now, and some ways in which fundamental theory contributes to the progress of specific theory in neoclassical positive economics will be discussed.

I will define a fundamental economic theory to be a body of propositions which describe, in general terms, the relation between institutional structure and individual behavior. This notion of fundamental theory is similar to the idea of equilibrium theory proposed by Daniel Hausman [5]. The most important difference is that, while Hausman conceives of equilibrium theory as a determinate, if implicit, body of propositions, I take fundamental theory to be a family of explicit theories which are distinct though closely related. For instance, one version of competitive theory might state that there are many firms with small efficient scales of production, while another might assume that each firm has a convex set of production possibilities but make no assumption about the number of firms. Each of these versions entails that the set of aggregate production possibilities for the economy is convex, which is an important condition for existence of equilibrium, and each covers some situations which the other does not.

As a rule, fundamental theories in neoclassical economics consider explicitly the optimization problems of all agents studied. That is, institutional structure is represented as a specification of these problems, and individual behavior of an agent is represented as optimal solution of a problem relative to some exogenously specified objective (i.e., tastes or profit maximization). Paradigmatic examples of fundamental theory are the theory of games (e.g., [9]) and the theory of general equilibrium (e.g., [3, 13]). In general equilibrium theory, for instance, consumers' demand (individual behavior) determines the market-clearing price (institutional structure), and the price determines consumers' budget sets in terms of which their demand is defined.

In contrast to specific theories, it is often difficult to state in concrete terms why a fundamental theory is of interest. For example, the response of a fundamental theorist to an inquiry about his particular field of study might be, "I study the difference between monetary and barter economies." This theorist presumably has no intention of conferring with an anthropologist when his research is done, to find out whether there is an example of a pure barter economy which refutes or corroborates his theory. Rather, he may be trying to formulate reasonable conjectures about functions called 'demand for money' and 'utility of money holdings' which appear in monetary theory at a more applied level. His hope is that, when the specific theory is augmented by these conjectures, it will be a better scientific theory in terms of predictive power, refutability, and so forth. Viewed in this light, the fundamental theorist's work seems much more oriented toward application than it did at first glance. His initial description of his work suggests a speculative concern with exotic or non-existent economies, but his actual concern is quite down to earth.

The foregoing example illustrates two facts about fundamental theory in economics. First, the distinction between fundamental theory and specific theory cannot be based on the practical orientation of the latter. It is not the case that fundamental theorists are interested in a different set of phenomena from their colleagues. Second, fundamental theory is not an alternative research program to specific theory. In particular, the formulation by fundamental theorists of formal systems (e.g., the Arrow–Debreu [3] model of general equilibrium) which differ markedly from those used in specific theory does not necessarily signal an effort to refute specific theory, or to compete with it in any way.

Fundamental theories may, as Hausman suggests in [5], serve an explanatory role of a special kind. However, this role does not exhaust their usefulness. In particular, fundamental theories can be used to help refine, test and evaluate specific theories. Six ways in which this may occur will now be enumerated.

First, fundamental theory in some instances extends the domain of application of specific theory. One example of this phenomenon is Arrow's [1] theory of state-contingent securities which has (among other applications) advanced the study of financial markets by providing a tractable representation of complicated and diverse portfolios of financial assets. Another example is Lancaster's [6] theory of hedonic pricing, which provides a basis for the study of markets in which there is substantial product differentiation (e.g., automobiles).

Second, as in the example of the introduction, fundamental theory sometimes suggests restrictions on the parameters of a specific theory which enhance the testability and predictive power of the latter. The best-known example of this is Slutsky's [11] characterization of demand functions which arise from utility-maximization. This characterization provides useful restrictions on the functional form of parametric demand functions for econometric estimation.

Third, fundamental theory provides a unified description of phenomena which are observed in diverse specific fields. For instance, there are convexity and continuity conditions which are sufficient both in game theory and in general equilibrium theory for existence of an equilibrium. If he can show that these conditions are met, a specific theorist is assured that he is working with a logically consistent set of assumptions. Moreover, in markets which behave pathologically (e.g., 'natural monopolies' where competition is technologically impossible to sustain), investigation of precisely how conditions for the existence and efficiency of competitive equilibrium fail is an informative diagnostic exercise. Such investigation sometimes leads to realization that apparently special features of diverse markets are in fact instances of a single phenomenon which can be systematically explained. A current attempt to unify such phenomena is the theory of signalling initiated by Spence [12]. This theory explains a number of instances of seemingly irrational market behavior (e.g., setting of educational requirements for job applicants which are unrelated to their prospective duties, payment by firms of dividends although earnings could be distributed by other means which receive more favorable tax treatment) in terms of the unobservability of quality differences among goods sold in a single market.

Fourth, fundamental theory reconciles specific theories which apparently conflict, by distinguishing between their domains of application. Typically, several specific theories are embedded in a single general theory in such a way that the original theories are recoverable by imposing special conditions on parameters. The classical case of such a reconciliation is Cournot's [2] embedding of the theories of monopoly and perfect competition in a game-theoretic theory of non-cooperative industry equilibrium. This general theory foreclosed the objection that refusal to consider industries with few firms as counterexamples to the theory of perfect competition, or refusal to consider industries with many firms as counterexamples to the theory of monopoly, constitutes an *ad hoc* avoidance of refutation.

Fifth, fundamental theory helps to isolate tests between conflicting specific theories by facilitating precise analytical comparisons between them. For example, there is a long-standing controversy in the applied theory of the firm concerning whether managers try to maximize profits or share value. Using the theory of state-contingent securities, it may be shown that these two objectives coincide for a manager if investors can use assets of other firms to hedge their investments in the firm in question. Armed with this result, empirical economists are able systematically to avoid trying to analyze data which could not possibly discriminate between the two hypotheses.

Sixth, fundamental theory provides informal criteria of coherence for specific theory. By this is meant something akin to the 'negative heuristic' described by Lakatos. Consider the case of Malthusian income-distribution theory. According to this theory, the working class population would grow until it was constrained by the imminent threat of famine, while capitalists would limit their family size in order to allocate some of their resources to saving and investment so that their consumption could be maintained at a comfortable level. This description of affairs was supposed to be true in the long run, without workers ever learning from observation of the capitalists how to better their lot. In order to express the Malthusian law within a theory of utility maximization, it must be assumed that workers possess an insatiable sexual appetite or a fanatical desire to have large families, while capitalists have epicurean tastes. Such an attribution of preferences to economic agents would be formally consistent, but it is incredible. As a matter of scientific common sense, this theory would be removed from consideration. If applied theory were formulated by placing hypotheses directly on the behavior of agents, the Malthusian theory would not be open to any objection on methodological grounds. This regulative use of utility-maximization seems to have an analogue in physics, where an empirically successful applied theory would be viewed with suspicion if intellectual contortions were necessary to reconcile it with entrenched conservation principles.

3. THE NATURE OF FUNDAMENTAL ECONOMIC THEORY

The discussion of the preceding section establishes that general equilibrium theory and game theory need not be viewed as economic theories which are in opposition to the less elaborate specific theory typically used in neoclassical economics, but that the former theories may be considered as giving an explicit foundation for the latter type. The present section is intended to support a stronger claim, that both types of theory are essential parts of any scientific research program in economics. This claim follows from two premises: First, that any complex of specific theories which is a competitor of neoclassical applied theory must also engender a fundamental theory which would serve the functions just described for the theories of utility-mazimizing agents. Second, that the complex of specific theories remains indispensable to the research program associated with the fundamental theory because the latter is not independently testable.

A general argument for the first premise will not be offered here. However, an example will be given which strongly suggests that a body of fundamental theory would arise within an alternative research program to neoclassicism. Consider a group of economists who accept the Malthusian population theory, and who are confronted with the rapid transfer of technology to third-world societies in the past several decades. This transfer indicates that groups of people who have traditionally enjoyed a low level of material welfare are willing and able to change their way of life radically in order to raise that level. This is *prima facie* evidence against the assumption that the working class will not curtail its birthrate to raise per capita income. In order to accommodate this evidence, the neo-Malthusians will require a fundamental theory (e.g., a detailed sociological theory of the influence of class on aspirations and behavior) which can explain why the entire population of a third-world country is not analogous to the working-class population of a technologically developed country. The reconciliation which this fundamental theory would provide between two specific theories (i.e., those of population growth and of diffusion of technology) would be the same sort of reconciliation as Cournot's theory provided between the specific theories of monopoly and perfect competition. Thus, as in the case of neoclassical economics, fundamental theory would arise in response to the needs of specific positive theory.

It might be supposed that, regardless of its origins, fundamental theory will ultimately become autonomous and challenge the role of specific theory in directing empirical inquiry. On the basis of such an assumption, fundamental economic theory has been judged to be an unsatisfactory scientific theory. In particular, it has been argued that either fundamental economic theory is presented at so high a level of abstraction that it is irrefutable, or else that its assumptions are so oversimplified that it must be false. If it is true that fundamental theory is autonomous, this argument may well be sound. However, the argument depends crucially on the assumption of autonomy. In the methodological writings of several economists there are two proposals to refute the argument (as applied to neoclassical theory) by denying the autonomy of fundamental theory. The remainder of this section will be devoted to characterizing the relation between fundamental and specific theory on which these proposals are based.

Friedman [4] and Machlup [7] have written standard expositions of one defense of the validity of utility maximization as a foundation for microeconomics. Both of these authors phrased their arguments in terms reminiscent of logical positivism, but that philosophical position is not essential to their point. What they suggested is that microeconomic theory is a partially interpreted theory, in which only qualitative propositions about market aggregates (i.e., price and quantity) are empirically significant. The consumers and firms whose optimizing decisions are studied in order to derive these propositions are to be regarded as purely theoretical entities, according to Friedman and Machlup, and are not to be confused with actual agents. Consequently, they claimed, neoclassical microeconomics cannot be criticized on the basis of evidence from sources such as surveys and psychological experiments, which show that actual consumers and firm managers sometimes make inefficient or inconsistent decisions.

While viewing exotic entities like quarks as fictitious may be a defensible position, it is not very persuasive to argue that the consumers and firms of economic theory bear only coincidental resemblance to actual ones. This dubious assertion is not necessary to do what Friedman and Machlup require, though. All that they need to assert is that neoclassical pure theory represents agents as maximizing utility only within a limited class of situations (i.e., those described in specific theories), and that failure of agents to maximize utility in situations outside that class (e.g., in experiments conducted in a psychology laboratory) would be irrelevant. On this view, an analogy might be drawn with the psychological theory that short-term and long-term memory are distinct mental processes. Most laboratory experiments would be taken to deal with short-term memory, and so could not refute a theory that agents have perfect recall of material in long-term memory. Similarly, market decisions are typically made by agents whose experience and opportunity for deliberation clearly distinguish their situation from that of subjects in artificial laboratory experiments. Thus, inefficient or inconsistent behavior by experimental subjects need not be counted as evidence against the hypothesis that economic decision-makers behave optimally.

On the view that economic decision-making is a distinguished sphere of behavior, economic phenomena provide the only reliable evidence about the hypothesis that such decision-making is rational. Fortunately, the domain of economic phenomena is large and diverse. Although fundamental theory must always be tested jointly with a specific theory, there are as many distinct tests as there are areas of specific theory. Specific theories, then, serve as auxiliary hypotheses or 'soft core theories' in tests of fundamental theory. Because no scientific theory is testable without such intermediation, pure economic theory is not in an exceptional position. In fact, fundamental economic theory may be particularly well situated because of the variety of tests offered by diverse specific theories.

Of course, the adoption of this view of the scientific status of fundamental theories does not make the immunity of neoclassical economics from psychological evidence a matter of *a priori* truth as does the partial-interpretation view. As Wilde [14] explains, the sphere of economic behavior might be construed to include laboratory situations. Whether this inclusion is appropriate is currently an unsettled question. If laboratory situations are included, then the theory of how subjects act is a specific economic theory, and a rejection of the utility-maximization hypothesis on the basis of experimental evidence would be consistent with the account just put forth of how fundamental theories are tested.

A substantially different defense of neoclassical fundamental theory was set forth by von Mises [8], who conceded that the hypothesis of utility-maximization is not capable of falsification, but who claimed that the theorists' knowledge that economic agents are utility maximizers is *a priori* and that concern about its falsifiability is inappropriate. A reasonable interpretation of these claims would be that fundamental theory is a logical meta-theory of specific theory, or that it is an explicit 'heuristic' of the scientific research program to which it belongs. On this view, the beneficial effects of fundamental theory discussed in the last section are not accidental 'spill-over' effects, but rather they constitute its essential scientific purpose. Fundamental theory is not really a theory at all, but instead it is a set of rules and recommendations about the construction and interpretation of applied theories. If the application of these rules and recommendations is fruitful, then fundamental theory is successful.

It is evident that both of the preceding accounts of fundamental theory presuppose the existence of a closely related body of specific theory. Although proponents of the two accounts might be seriously divided on other issues, they would agree that the critics of fundamental theory are attacking a straw man. Fundamental theory simply does not exist as a self-contained body of putative scientific knowledge.

4. SUMMARY AND CONCLUSIONS

Terms like 'competing research program' and 'problem-shift' have sometimes been used to describe the emergence of an explicit body of fundamental theory alongside the specific theory of neoclassical economics. Such descriptions imply that fundamental and specific theory conflict or are incommensurable. The argument of the present paper began with a detailed account of some of the close relations between fundamental and specific neoclassical economic theory. The existence of these relations indicates that conflict or incommensurability between the two types of theory is a much less serious problem than might have been thought, and it suggests that a different description of how the two types are related might be appropriate.

The argument continued by suggesting that the emergence of fundamental theory is a characteristic event in the evolution of research programs in economics, rather than being a peculiar event in the history of neoclassical economics. The description is clearly consistent with the occurrence of the kinds of interactions between fundamental and specific theory which were already noted. Furthermore, the account of fundamental economic theory given here facilitates a clear statement of the replies of several prominent economists to objections raised against it. Specifically, the account enables these replies to be stated without appeal to idiosyncratic or contested philosophical theories of knowledge and meaning. While the arguments considered were originally put forth in terms of the neoclassical theory of utility maximization, they have been shown here to be instances of general views about the relation between fundamental and specific theory in any economic research program. While they differ on what the role of fundamental theory is, each of these general views is clearly analogous to a standard account of the nature of physical sciences.

While it is hoped that the present discussion has been sufficiently detailed to inspire confidence that fundamental economic theory is a legitimate scientific endeavor, obviously it has been far from complete. To begin with, each of the examples given in Section 2 deserves a much more detailed examination than it has received. Also, it would be of interest to have corresponding examples from other economic research programs besides the neoclassical one. A striking feature of the definition of fundamental theory used here, having implications which deserve careful attention, is that it places far more emphasis on institutions than is traditional. An explicit theory of the domain of economic phenomena and of economic evidence is needed if the view that fundamental theories are tested via specific theories is to be considered seriously. The view of von Mises also raises questions, for instance, about how the deductive arguments of fundamental theories contribute to their usefulness as prescriptions. Work on problems such as these should lead to a clearer understanding of the structure of economic reasoning, and of the relation between economics and the other sciences.

California Institute of Technology

NOTE

* I am indebted to Daniel Hausman and Steven Pollack who provided detailed and valuable comments on an earlier draft of this paper.

REFERENCES

[1] Arrow, K., 'The role of securities in the optimal allocation of risk-bearing', *Review of Economic Studies* **31** (1963–1964), 91–96.

[2] Cournot, A., *Researches into the Mathematical Principles of the Theory of Wealth*, 1838, reprinted 1963, Homewood, Ill.: R. D. Irwin.

[3] Debreu, G., *Theory of Value; an Axiomatic Analysis of Economic Equilibrium*, New Haven: Yale University Press, 1959.

[4] Friedman, M., 'The methodology of positive economics', in *Essays in Positive Economics*, Chicago: University of Chicago Press, 1953.

[5] Hausman, D., 'Does general equilibrium theory explain anything?' this volume, pp. 17–32.

[6] Lancaster, K., *Consumer Demand; A New Approach*, New York: Columbia University Press, 1971.

[7] Machlup, F., 'Theories of the firm: marginalist, behavioral, managerial', *American Economic Review* **57** (1967), 1–33.

[8] Mises, L. von, *Human Action*, New Haven: Yale University Press, 1949.

[9] Neumann, J. von and O. Morgenstern, *Theory of Games and Economic Behavior*, 1944, reprinted 1967, New York: Wiley.

[10] Silberberg, E., *The Structure of Economics*, New York: Academic Press, 1978.

[11] Slutsky, E., 'On the theory of the budget of the consumer', [1912] in G. J. Stigler and K. E. Boulding (ed.), *Readings in Price Theory*, Homewood, Ill.: R. D. Irwin, 1952.

[12] Spence, A. M., *Market Signalling; Informational Transfer in Hiring and Related Screening Processes*, Cambridge: Harvard University Press, 1974.

[13] Walras, L., *Elements of Pure Economics, 1874–1877*, reprinted 1965. Homewood, Ill.: R. D. Irwin.

[14] Wilde, L. 'On the use of laboratory experiments in economics', this volume, pp. 137–148.

DANIEL M. HAUSMAN

ARE GENERAL EQUILIBRIUM THEORIES EXPLANATORY?

The elegant theories of general equilibrium which have been developed during the past three decades leave many economists puzzled, since they appear to have so little to do with real economies.[0] Gerard Debreu in his classic *Theory of Value* states that his theory is concerned with the explanation of prices.[1] Yet others as distinguished as Kenneth Arrow and Frank Hahn deny that general equilibrium theories are explanatory.[2] In this paper I shall offer a philosophical interpretation of what these mathematical structures called general equilibrium theories are. I shall defend their cognitive worth, although I shall argue that they are virtually without explanatory power.

To understand what general equilibrium theories are, one must first understand what *equilibrium theory* is. General equilibrium theories are applications of (and thus *not* identical to) equilibrium theory. They are, as will be discussed later, the result of combining equilibrium theory with auxiliary hypotheses of the right sort. 'Equilibrium theory' is my name for the fundamental theory of microeconomics. Most, although not all, modern economists believe that equilibrium theory has a good claim to be considered as *the* fundamental economic theory. It is the theory to which they hope to reduce, or at least relate, macroeconomic theories. It is the theory which they hope can be extended to deal with questions of economic growth and change. It is the theory which the economist relies on in empirical research and in many welfare recommendations. When one has succeeded in saying clearly what equilibrium theory is, one has largely succeeded in saying what neo-classical economics is. The theory may sensibly be called 'equilibrium theory', since it is concerned with the factors that influence economic equilibria. An economic equilibrium is a state of affairs in which there is no excess demand – a state of affairs in which at the going prices nobody wants to go on exchanging. Note that I am claiming that equilibrium theory, *not* general equilibrium theory, is the fundamental theory of contemporary economics.[3]

What is equilibrium theory? It may be formulated roughly as the following eight lawlike statements. Qualifications are needed and will be discussed below.

J. C. Pitt (ed.), Philosophy in Economics, 17–32.

(1) For any individual A and any two options x and y, one and only one of the following is true: A prefers x to y, A prefers y to x, A is indifferent between x and y.

(2) A's preferences among options are transitive.

(3) A seeks to maximize his or her utility where the utility of an option x is greater than the utility of an option y for A if and only if A prefers x to y. The utilities of options are equal just in case the agent is indifferent between them.

(4) If option x is acquiring commodity bundle x' and option y is acquiring commodity bundle y' and y' contains at least as much of each commodity as x' and more of at least one commodity, then all agents prefer y to x.

(5) The marginal utility of a commodity c to an agent A is a decreasing function of the quantity of c that A has. The second partial derivatives of the utility function, considered now as a function mapping commodity bundles to real numbers, are negative.[4]

(6) When we increase any input into production, other things being equal, output increases, but, after a certain point, at a decreasing rate.

(7) Increasing all the inputs into production in the same proportion increases output by that proportion. The production set is non-concave.

(8) Entrepreneurs or firms attempt to maximize their profits.

Many would say that (1)–(3) assert that people are rational and that (1)–(4) assert that people are economically rational.[5] Notice that (4) has the effect of identifying options and commodity bundles. Diminishing marginal utility is supposed to be a psychological law of some sort, which is sometimes (implausibly in my opinion) regarded as part of what it means to be economically rational. Constant returns to scale and diminishing returns to variable inputs are supposed to be in some sense natural laws or technological givens.[6] To say that firms attempt to maximize profits is supposed to be a lawlike claim concerning motivation, possibly deducible from other features of our theory.[7]

This set of lawlike statements is disquieting, since, even with qualifications, most of the claims appear to be false.[8] Others seem reliable only in certain circumstances, given a certain institutional setting. Presumably, for example, one of the factors that makes it reasonable to claim that individuals generally prefer more commodities to fewer, (4), is that there is a market on which excess commodities can be easily traded.[9] (1)–(8) do not, of course, exhaust the general 'laws' of microeconomics. (1)–(8) (or some reformulation of them) are, however, the basic 'laws'. Other important lawlike claims, such as the claim that the price of a commodity will rise when demand for it exceeds its supply, can be derived from (1)–(8) and various stipulations concerning markets, information, and the like. The above eight lawlike statements express the fundamentals of neo-classical economics.

To assert that equilibrium theory is well formulated by (1)–(8) is rough, misleading and incomplete. It is rough because (1)–(8) are not a very precise statement of the theory. It is misleading, because neo-classical economists do not always make use of *all* of the above lawlike statements. (7) and (8) are the most likely to be dropped or replaced by contrary generalizations, while revealed preference theory is supposed to supplant (1)–(5). Formulating equilibrium theory as (1)–(8) is also incomplete. To understand what equilibrium theory is, one must also make reference to commonly used mathematical techniques and standard characterizations of the institutional, informational, and material background. These characterizations may themselves be lawlike.[10] They differ, however, in three ways from (1)–(8) above. First, they have a narrower scope. The assertion that there are many buyers and sellers in every market is not supposed to be true of all or nearly all societies. Second, these characterizations of prevailing circumstances are not regarded by economists as discovered by or as asserted by their theory. They provide a sketch of the circumstances to which the theory is applied and are thereby crucial to the derivation of important theorems, but they are not themselves assertions of the theory. The term, 'assumption', of which economists are perhaps overly fond,[11] fits these stipulations well. Finally, although these claims concerning markets, resources, information and the like are important to equilibrium theory, they are less important to the identity of the theory than are (1)–(8) or some reformulation of them. Neo-classical economic analyses generally assume, for example, that commodities are infinitely divisible. This assumption has, as the result of employing better mathematical techniques, become avoidable in certain circumstances. Were it to become fully dispensable, economists would not, I believe, conclude that there had been any great revolution in microeconomics. If, however, a number of our mathematical techniques and assumptions concerning prevailing circumstances were to change, we might well regard our fundamental theory as changed.[12] Replacing any one of (1)–(8) with a non-equivalent generalization would, on the other hand, count by itself as a theory change. The distinction between the fundamental laws and the generalizations about prevailing circumstances is not, however, quite as sharp as the above discussion suggests. (7), for example, is a borderline case.[13] Most economists would regard it as an 'assumption' rather than as an assertion of the theory, yet its scope os supposed to include all economies. It is not nearly as crucial to the identity of equilibrium theory as is, for example, (3).

Many economists would regard the above discussion as misconceived, because they believe (mistakenly, in my opinion) that economic theorists are not concerned to discover laws and theories, but to fashion models that serve some explanatory, predictive or theoretical purpose.[14] (1)–(8) should be regarded as assumptions needed in the construction of various models, not as lawlike assertions of a theory. The issues involved in this objection are serious and complicated. It is certainly true that economists think in terms of models. The question is whether models are alternatives to theories which avoid employing lawlike generalizations or whether models are rather a particular kind of theory. My view, in outline, is that at least in so far as economists are concerned to *explain*, they cannot avoid including lawlike generalizations in their models. Without establishing reliable generalizations, their models will have little predictive,[15] let alone explanatory worth. Without laws or something like laws, how can economic models help us to explain? This sketchy response is, of course, hardly the last word on this difficult issue.

With this analysis of equilibrium theory in mind, we can now see what general equilibrium theories are. They are applications of equilibrium theory – combinations of (1)–(8) above with further auxiliary hypotheses. These further auxiliary hypotheses are of two varieties. First one needs to make numerous stipulations concerning information, market structure, the divisibility of commodities, etc. These stipulations are similar to (indeed often identical to) those which are made when one applies equilibrium theory in other and less problematic ways. Partial equilibrium analyses make much the same stipulations.[16] Second, general equilibrium theories, unlike other applications of (1)–(8), assert or assume that there are many commodities and that there is a general interdependence among the various markets. Assumptions of this second sort are what distinguish *general* equilibrium theories.

There are two quite different varieties of general equilibrium theories which need to be sharply distinguished. One of these is of practical use, while the other is quite abstract. The first kind is exemplified by input – output models. By assuming, for example, that there are constant production co-efficients and that demand will show special constancies, we can set up a model of an economy with perhaps a hundred different commodities and industries and, with the help of a computer, investigate how it operates. Practical general equilibrium theories raise no problems about explanation that do not arise equally with respect to partial equilibrium analyses.

Theories of the second kind, which I shall call 'abstract general equilibrium theories', place no limitations on the interdependence of markets or on the nature of production and demand beyond those implicit in our lawlike claims. When economists speak of general equilibrium theory, it is usually this second, abstract variety that they have in mind. It is abstract general equilibrium theory with which I am concerned in this paper. Given the abstractness and lack of specification in abstract general equilibrium theory, many economists regard it as the fundamental theory of contemporary economics. As the previous discussion suggests, this seems to me a mistake. *Equilibrium theory* is the fundamental theory. General equilibrium theory is a particular application of the fundamental theory. This is not an arbitrary stipulation on my part. When scientists have some explicit explanatory or predictive purposes or when they want to develop a theory of a specific subject matter, they often *combine* some fundamental theory with further auxillary hypotheses. Such combinations are precisely what general equilibrium theories are. They are thus unlike fundamental theories in the natural sciences. Equilibrium theory, on the other hand, serves much the same functions in economics that fundamental theories in the natural sciences serve in their respective domains.

What confuses matters is that applying equilibrium theory as the general equilibrium theorists do serves no clear explanatory or predictive purposes. Nor are these theorists attempting to develop a theory of a more specific subject matter within economics. The stipulations they make concerning information, markets and the like are ill-suited for any such purposes. Theories of intertemporal general equilibrium assert or assume that agents have complete and accurate knowledge concerning the availability and prices of commodities and concerning the production possibilities both in the present and the future. They also stipulate that there is a complete set of commodity futures markets on which present commodities (or titles to future commodities) can be freely exchanged for titles to future commodities of every kind and date.[17] Since such claims render the theory so obviously either false or inapplicable to real economies, little testing can be done.[18]

Furthermore, the fact that we cannot satisfy, even approximately, such assumptions of the theories leaves abstract general equilibrium theories with little if any predictive worth.[19] Given the falsity of stipulations such as perfect information, one wants to know what the point is of abstract general equilibrium theories.

One further peculiarity of abstract general equilibrium theories is that they almost invariably take the form of existence proofs. One demonstrates that the axioms (which include reformulations of (1)–(8) and stipulations or auxiliary hypotheses of the kinds discussed) are sufficient conditions for the existence of an economic equilibrium. Abstract general equilibrium theories thus seem to have the form of explanatory arguments where the explanandum is the existence of an economic equilibrium. Yet construing general equilibrium theories as explanations of economic equilibria with various properties is implausible, since no economy is ever in a state of economic equilibrium. There is no fact of equilibrium to be explained.[20] Are such peculiar theories thus without explanatory power? If so, what value can such abstract applications of equilibrium theory have?

Answering the first of the above two questions requires saying something about explanation. The much battered starting point for modern discussions of explanation is Carl Hempel's deductive nomological model.[21] A necessary condition for giving a scientific explanation for some event or state of affairs supposedly is deducing a description of that event or state of affairs from a set of true statements, some of them lawlike. Sufficient conditions are much more difficult to give and are supposed to involve pragmatic considerations. Nowadays most philosophers of science believe that the deductive nomological model is too restrictive. Not only do there appear to be statistical explanations, but, in fact, explanations of events or states of affairs in the sciences are rarely, if ever, deductive nomological explanations. To deny that general equilibrium theories are explanatory on the grounds that they do not give us deductive nomological explanations of the characteristics of real economics is unreasonable.

Consider what happens when natural scientists attempt to deal with complex real systems. Take the trite example of the path of a leaf's fall. In what sense can it be explained by our physics? Precise deductive nomological explanation seems out of the question, because we cannot get precise information concerning all the relevant initial conditions and because of the mathematical difficulties of calculating the path, even if we did have all the relevant information. If these problems of knowing the initial conditions and of calculation were our only problem, then we would have what I would call 'an explanation in principle' of the leaf's path.[22] We would know all the relevant laws and causal factors. Even though we could not give an actual explanation in detail of the path, there would be nothing mysterious about falling leaves and nothing general or theoretical to learn about them.

We bring ourselves to the point not only of being able to give explanations in principle, but also of believing that we are able to give them, by developing laws and theories that we test in other, simpler circumstances. In the case of the falling leaf, we test our theory of gravitation on bodies falling in evacuated chambers or by seeing how well it accounts for the motion of the planets. Our theories of fluids and of resistance are also developed and tested with respect to simpler situations. The difficulty comes in convincing ourselves that we know *all* of the laws which are relevant to the case of the falling leaf. We cannot know for certain that we have learned about all the relevant forces – perhaps telekinesis is a significant phenomenon with leaves – but we can have good reason to believe that we have learned all the general truths needed to explain in principle the leaf's path. We have such good reason when with simplifications concerning initial conditions we can make roughly correct predictions concerning falling leaves and are able to cite the factors responsible for any appreciable errors. Achieving a theory of falling leaves in this way is employing what J. S. Mill called the deductive method.[23] Mill believed that our economic laws are established inductively by psychology and the natural sciences and that we develop our economic theory deductively.[24]

Even such explanation-in-principle is a very tall order. I doubt that we can now explain in principle the path of a falling leaf and we certainly can not explain in principle either local or overall characteristics of any real economy. Not only do we encounter great difficulties confirming our purported laws – perhaps because of the virtual impossibility of setting up simplified experimental situations – but we know full well that we are ignorant of many relevant laws and that we are leaving out of account many significant causal factors.[25] At best we believe that we have got what Mill calls an 'inexact science' – an account of the principal causal factors involved.

Inexact sciences certainly appear to be explanatory. We believe that physicists can to some extent explain the paths of falling leaves even if they cannot give a full explanation in principle, The theory of gravitation explains many of the characteristics of tides, even though we remain ignorant of what Mill calls the 'minor causes'.[26] It is difficult to give a satisfactory analysis of this "inexact explanation in principle," but I believe that equilibrium theory provides just this sort of explanation for the fact that, for example, the price of coffee increased after frost destroyed the Brazilian coffee harvest.[27]

These features of explanations can be regarded as two difficulties with the deductive nomological model. First, the inexactness or incompleteness of our theories can be captured formally by the conclusion, rightly stressed by Alexander Rosenberg,[28] that scientific laws invariably carry implicit or explicit ceteris paribus clauses.[29] These gum up the working of the deductive nomological model by presenting us with a choice among three alternatives. First we might simply ignore the ceteris paribus clauses. The result is that our lawlike claims are no longer true (or well-confirmed), and we thus violate one of the conditions of the deductive nomological model. Second, we might include the ceteris paribus clauses, in which case we can no longer deduce the explanandum without the additional stipulation that *ceteris* is *paribus*. We will often lack good reason to defend such an assertion. Moreover such an assertion will rarely if ever by precisely true. Third, we might conclude that we can only deduce a weaker explanandum itself qualified with a ceteris paribus clause.[30] To select this last option is to abandon the ambition of explaining actual events or states of affairs.[31]

The other problem, the difficulty of precisely specifying the initial conditions and then calculating the implications turns up as the fact that the auxiliary hypotheses or statements of initial conditions which we actually employ in our explanatory arguments will include some that are known to be false. We explain the orbits of the planets using Newtonian theory by denying that the planets attract one another and then correcting for the error. We use Newton's laws to explain the path of disk of solid CO_2 moving on a smooth surface by asserting that the force of friction is zero. Such ordinary explanations violate Hempel's requirement that the statements of the explanans be true (or at least well-confirmed).

The above difficulties do not constitute any knock-down refutation of the deductive nomological model. There are formal devices by means of which one can cling to the deductive nomological model while conceding that the best we can usually manage in science is inexact explanation in principle. I doubt, however, that such defenses are worth making. Although I agree with Hempel that pointing out nomic connections is central to explaining, the sciences do not give us the sort of grip on particular events and states of affairs that the deductive nomological model suggests. This fact should be clarified, rather than concealed by philosophical models of explanation.

I am not yet prepared to offer my own model of explanation, although I shall argue for at least one necessary condition that scientific explanations satisfy, but which general equilibrium theory does not. That necessary condition on explanations is the following:

(9) All statements of initial conditions or auxiliary hypotheses must either be true or approximately true.

(a) The falsity of those which are not true must sometimes be of little significance.

(b) We must have some knowledge about when the falsity of our statements of initial conditions or auxiliary hypotheses will lead us astray.

(a) and (b) are further necessary conditions which statements of initial conditions must satisfy in order to be considered as approximately true.[32] In sanctioning the use of idealizations, one must not wind up disregarding all questions about the truth or empirical adequacy of the statements in our purported explanations. (9) is the sort of necessary condition which would fit into a model of explanatory arguments. Philosophers like Wesley Salmon have argued persuasively against any such models.[33] While these issues demand further discussion, it seems to me most appropriate to assess general equilibrium theories with respect to models of explanatory arguments, since, as existence proofs, they *are* arguments. Moreover, employing a relevance model like Salmon's will not change one's verdict on general equilibrium theory.[34] One might object to the whole idea of formulating a necessary condition like (9).[35] Even if it is sensible to formulate a timeless 'logic' of scientific explanation like Hempel's, surely nothing as loose and messy as (9) can capture a necessary condition on scientific explanations in general. At best (9) gives a simplistic account of features which are currently important in scientific explanations. This criticism has merit, but it in fact challenges nothing about (9) except its superficiality. Unless one believes that general equilibrium theories must be considered explanatory because many economists think that they are, one has no reason to object to generalizing about features of what are now considered to be explanations in the natural sciences and to inquiring whether the existence proofs of general equilibrium theories share these features.[36] Of course, I may be mistaken in suggesting that (9) is a reasonable necessary condition on scientific explanation. If I am, my mistake reveals a misapprehension concerning the characteristics of actual scientific explanations, not an inadequacy in my methodology.

Even a weak necessary condition like (9) cannot be satisfied by current general equilibrium theories.[37] The problem I am concerned with does not involve the 'laws' of general equilibrium theory. As mentioned before, equilibrium theory in its less high flown applications appears to be explanatory.[38] The difficulty resides in the particular auxiliary hypotheses that general equilibrium theories include. It is not the case that we can show that it is true or approximately true that individuals are omniscient about the future or in virtually any case that their lack of omniscience is of little significance.[39] Condition (9) is closely connected to the testability of a theory; untestable theories will not satisfy it. Yet it is not, contrary to Milton Fiedman's views,[40] equivalent to the demand that the implications of our theory be confirmed. If we must thus conclude that general equilibrium theories do not explain the characteristics of real economies, what worth do they have? Is the work of the abstract general equilibrium theorists more than some pretty mathematics? Can abstract general equilibrium theories be of value if they have nothing to tell us of real economies?

These are difficult questions upon which leading theorists disagree. Some believe (mistakenly, if the above argument is correct) that general equilibrium theories serve at least in part to explain prices.[41] C. J. Bliss denies that abstract general equilibrium theories 'represent reality', but claims that nevertheless they are a good point of departure and a good guide to which concepts are central and fundamental.[42] Although Bliss's view suggests important truths, it is misleading. Many economists, particularly when they are concerned about how to justify their theories, are tempted to say that they only provide us with some sort of logic of economic phenomena or that they are merely bags of tools into which we dip when convenient.[43] These claims have a certain truth to them, which I have tried to capture by distinguishing equilibrium theory from its applications. They do not, however, resolve problems of justification. If an economic theory is only a logic or a bag of tools or a guide to which concepts are central, we still need to ask whether it is a good logic or a good bag of tools or a good guide. If, as in the case of general equilibrium theories, we cannot apply the theory, we have no way of answering these questions.[44]

There is, however, more to the attitude toward general equilibrium theory that Bliss and others hold than the above argument recognizes. General equilibrium theory may be of great heuristic value.[45] Although heuristics is itself a complicated subject, one can show that general equilibrium theories have been of heuristic value merely by showing (as one easily can) that they have in fact helped in developing valuable empirical economic theories. Notice that the heuristic value of general equilibrium theories is independent of the existence proofs (the arguments) that such theories provide.[46] Where the general equilibrium theories have been most valuable has been in the invention of conceptual and mathematical devices (dated commodities, for example) which are useful in other theories.

Yet it seems to me that the existence proofs that general equilibrium theories provide are themselves also of value. Frank Hahn and Kenneth Arrow largely deny that they have anything to tell us about real economies, but they insist, rather unclearly, that they remain a serious and valuable part of economics.[47] My view is largely a development and elucidation of Arrow and Hahn's position.[48]

Since the Eighteenth Century many economists have believed that, given reasonably favorable conditions, self-interested voluntary exchanges lead to coherent and efficient economic organization. Yet the theories which economists have possessed have not enabled them to explain how this order in fact comes about nor even to show how it is possible that such order could come about. Economic theorists might thus reasonably be in doubt concerning both whether their theoretical framework captures the crucial features of the economy and whether it is likely to lead them to an adequate theory.[49] In pursuing and developing equilibrium theory, will one ever be able to explain how self-interested individual action within certain institutional constraints can lead to coherent economic order? Do we really have a grip on the most important and central economic regularities? Will we ever be able to understand whether the results of individual actions are truly efficient and whether they lead to the achievement of other goals we might have?

Clearly there is no way to show in advance that equilibrium theory, when refined and developed, will ever get us close to being able to explain in principle detailed as well as overall characteristics of economies or of economic events. The successes of partial equilibrium analyses or of practical general equilibrium models give us some reason for hope. Furthermore, both as a way of working toward such an adequate theory and as a way of testing the promise and power of the equilibrium theory that we possess, it is sensible and important to ask whether (1)–(8) are, when supplemented by further stipulations about background conditions, sufficient to explain the existence of any sort of economic equilibrium, even if it is only imaginary. Abstract general equilibrium theorists impose the additional condition that the stipulations or auxiliary hypotheses concerning prevailing circumstances be of the same general kind as those commonly employed in practical empirical applications of (1)–(8). One excludes here, of course, those stipulations that restrict the generality of the equilibrium investigated. If (1)–(8)

and these additional stipulations in some sense explain how economic order *can* result from independent self-interested individual actions, we have reason to believe that we have captured the fundamental elements of economic life and that our theory is the right way to proceed.

In proving the existence of equilibria under various conditions, I take the abstract general equilibrium theorists to be providing us with explanations in principle of the characteristics of possible (although imaginary) economic states. In doing so they demonstrate to us that equilibrium theory is capable of explaining at least some sorts of complicated economic equilibria, and thus they give us reason to believe that we are on the track of an adequate general economic theory. This sort of an explanation of a possibility needs to be distinguished carefully both from explaining 'How possibly?' in the sense of Hempel and Dray and from any discussions of the feasibility of economic equilibria. Hempel's and Dray's view is that sometimes things happen contrary to our expectations which need explaining (away).[50] But economists are not trying to explain away the existence of some sort of competitive equilibrium, to show that its existence is consistent with prior beliefs. Nor, despite misleading claims by Frank Hahn,[51] are abstract general equilibrium theorists concerned with how or whether a competitive equilibrium is practically possible or feasible. We do not need all this theory to know that real semi-competitive capitalism does not regularly achieve full employment. If we did need general equilibrium theories for the purpose, they would not help us anyway, since the existence proofs that the theories provide show us only what conditions are sufficient for competitive equilibria, not what conditions are necessary.

The abstract general equilibrium theorists have shown us that were the world very much simpler than it is or will ever be, we could use our laws to explain in principle how economies work. If one regards the resemblances between the imaginary worlds of the theories and actual economies as at all significant,[52] these demonstrations give us reason to believe, in Mill's words, that we know the laws of the "greater causes"[53] of economic phenomena. We thus have reason to believe that we are on the right track. We should regard the existence proofs as providing this sort of theoretical reassurance, not as explanations. Note in addition that these abstract general equilibrium theories may help us to improve our current economics. By progressively weakening and complicating the stipulations needed in order to demonstrate the existence of more complex equilibria, we come closer to being able to apply the theory to real economies.

There is, however, only so much that mathematical work can do. So long as the set of lawlike statements, (1)–(8) above, remain rough and dubious, economic theory can hardly match the standard that physical theory sets.[54] Problems with testing and justification are serious. Much difficult work remains to be done in providing a general characterization of the sort of economic order (and disorder) there actually is that needs explaining. Yet the mathematical efforts of the abstract general equilibrium theorists might even help us out with these difficulties. They might, for example, show us how to make do with weaker formulations of (1)–(8) or suggest to us replacements. In the meanwhile, we can return to our practical applications of equilibrium theory with somewhat more confidence that we do know the "greater causes" of economic phenomena.

University of Maryland

NOTES

[0] They have puzzled me, too. In resolving that puzzlement many people have helped me. Special thanks go to Lindley Darden, Philip Ehrlich, Ronald Findlay, Catherine Kautsky, Eva Kittay, Isaac Levi, Sidney Morgenbesser, Frederick Suppe and the participants in the Philosophical Problems of Economics Workshop, especially Edward Green and Louis Wilde.

[1] (New Haven: Yale University Press, 1959), p. ix.

[2] *General Competitive Analysis* (San Francisco: Holden Day, 1971), pp. vi–viii.

[3] I am here excluding Marxian economics and recent innovations like neo-Ricardian theory, not because they are uninteresting, but because their impact on mainstream economics has thus far been small.

[4] As everybody learns in their first economics course we need not attach significance to utility differences. We thus learn to speak of consumers' marginal rates of substitution rather than their marginal utilities, but nothing is lost by speaking roughly here.

[5] See my review of Alexander Rosenberg's 'Microeconomic laws: a philosophical analysis in *Nous* **13** (1979), 120–121.

[6] See for example Alexander Rosenberg, *Microeconomic Laws: A Philosophical Analysis* (Pittsburgh: Pittsburgh University Press, 1976), pp. 29–30. His discussion is somewhat confusing (as is this whole interpretation of (6) and (7)) because the production function for steel, for example, unlike a chemical formula, makes, among other things, an assertion about the state of our *knowledge*.

[7] If I saw how this deduction could be carried through, I would drop (8) from the list. (8) is particularly problematic because entrepreneurs are also consumers and (8) may thus conflict with (3).

[8] This assertion is not my discovery. Most economists would concede that (1)–(8) are not true. A few moments of reflection by the reader will convince him or her that a number of these 'laws' are at the very least in need of lots of qualifications.

[9] But (4) is not derivable from (1)–(3) and (5)–(8) conjoined with auxiliary hypotheses

concerning the prevailing circumstances. There are, of course, alternatives to (4) which allow us to derive (4) as a theorem, but none of these alternatives is appreciably weaker or more reasonable than (4) itself.

[10] Given the enormous philosophical difficulties concerning the nature of laws, I do not want to stake my contrast between (1)–(8) and other claims which are prominent in equilibrium theory on the (greater) lawlikeness of (1)–(8).

[11] A whole controversy on 'assumptions in economics' followed the publication of Milton Friedman's, 'The Methodology of Positive Economics', in *Essays in Positive Economics* (Chicago: University of Chicago Press, 1953), pp. 3–42. Friedman calls all the axioms of a theory its 'assumptions'. The term has been used by others in a number of different ways. See, for example, K. Brunner, '"Assumptions" and the cognitive quality of theories', *Synthese* **20** (1969), 501–525.

[12] Yet nineteenth-century 'classical mechanics' is still considered to be Newton's theory of mechanics.

[13] Paul Samuelson argues that economists use (7) as if it were a logical truth. See his *Foundations of Economic Analysis* (Cambridge, Massachusetts: Harvard University Press, 1947), p. 84.

[14] Hal Varian and Alan Gibbard in their recent 'Economic models', *Journal of Philosophy* 75 (1978), 664–677 came close to espousing such a position. It is particularly popular among empirical economists.

[15] Simple induction need not invoke any laws, but I would argue that any extension of our abilities to predict to new circumstances relies at least implicitly on generalizations.

[16] Partial equilibrium theories or analyses are applications of equilibrium theory in which one employs certain specific kinds of auxiliary hypotheses. These are of two distinctive kinds. First, certain economic magnitudes like prices, incomes, quantities produced – which are in fact influenced by the workings of the market and which are sensitive to other factors in the economy – are taken as constants whose values are given. This permits independent consideration of single markets and permits one to consider factors influencing supply separately from those influencing demand. Supply and demand are principally partial equilibrium concepts. Second, to simplify matters, one works with certain sorts of aggregates. Instead of considering the complex changes in prices that result when one employs different techniques of production, one might assume that all consumption goods can be treated as one commodity and all capital goods as another single commodity. If one is concerned with certain aspects of consumer choice, one might assume that all but one or two commodities which are of particular interest can be considered as a single composite commodity. Notice that it is incorrect to assert that microeconomics never deals with aggregates.

These isolating and aggregating auxiliary hypotheses distinguish partial equilibrium applications of equilibrium theory. In each case the state of mutual adjustment or equilibrium we are considering is only partial, since we are ruling out adjustments which we know would have to occur in other markets or in specific outputs or prices.

[17] For good expositions see T. C. Koopmans, 'Allocation of resources and the price system', in *Three Essays on the State of Economic Science* (New York: McGraw-Hill, 1957), esp. pp. 105–126; E. Malinvaud, *Lectures on Microeconomic Theory*, tr. A. Silvey (Amsterdam: North-Holland, 1972); and C. J. Bliss, *Capital Theory and the Distribution of Income* (Amsterdam: North-Holland, 1975), Ch. 3.

[18] Indeed the whole notion of testing general equilibrium theories is somewhat strange since, as discussed below, they are presented in the form of proofs.

[19] Thus Milton Friedman, quite consistently, is unsympathetic to the 'Walrasian viewpoint'. See 'The Marshallian demand curve', in *Essays in Positive Economics*, pp. 89–92.

[20] Qualifications are needed. There are special kinds of equilibria that portions of economies occasionally approximate. I am thus not denying that on rare instances general equilibrium theories may be explanatory. All I am insisting on is that these instances are rare and quite special. If these theories have any appreciable importance, which I believe they do, we must thus look for other virtues than their apparent ability to explain under peculiar circumstances.

[21] The most refined presentation is in 'Aspects of scientific explanation', in *Aspects of Scientific Explanation and Other Essays in the Philosophy of Science* (New York: Free Press, 1965).

[22] Note the subjunctive. I am not asserting that physicists can currently offer an adequate explanation in principle of the path of a leaf's fall.

[23] See *A System of Logic* (1843; rpt. London: Longmans, Green & Co., 1949), Bk. III, Ch. XI and Bk. VI, Ch. IX.

[24] See his *Logic*, esp. Bk. VI, Ch. IX, Sec. 3. See also J. E. Cairnes, *The Character and Logical Method of Political Economy* (2nd Ed. 1888; rpt. New York: A. M. Kelley, 1965), pp. 71f.

[25] No unfavorable comparison between economics and the natural sciences is here intended.

[26] *Logic*, Bk. VI, Ch. III, Sec. 1.

[27] Notice that in this paper I am not challenging the explanatory power of equilibrium theory. There are difficulties about considering equilibrium theory to be explanatory which result from the roughness of its 'laws'. I discuss these in an unpublished paper 'Defending microeconomic theory', delivered at the Michigan State Conference on Philosophy and Economics, May, 1979.

[28] *Microeconomic Laws*, esp. pp. 134–137, 110–112.

[29] The link between inexactness and ceteris paribus clauses requires some defense. I argue for it in 'Economic models: a philosophical inquiry into capital theory', diss. Columbia University, May, 1978, pp. 240–244.

[30] This possibility was suggested to me by Eva Kittay. It is more interesting with respect to the explanation of laws.

[31] Under a description, of course. Abandoning this ambition may in the end turn out to be reasonable. I'm not sure.

[32] (a) and (b) were suggested to me by rough comments of Terence Hutchison in *The Significance and Basic Postulates of Economics* (1938; rpt. with a new preface. New York, A. M. Kelley, 1960), esp. p. 46.

[33] See his 'Statistical explanation', in *The Nature and Function of Scientific Theories*, ed. R. Colodny (Pittsburgh: Pittsburgh University Press, 1970), pp. 173–232.

[34] General equilibrium theories will generally not identify or cite any statistically relevant factors that more plebian applications of equilibrium theory overlook.

[35] I have benefitted here from conversations with Philip Ehrlich.

[36] I am here ignoring possible anti-naturalist scruples against doing so.

[37] This is not intended as a criticism of general equilibrium theories. Why must all valuable parts of science be explanatory? If I am here criticizing anything, my object

is the misinterpreting of general equilibrium theory of which some economists have been guilty.

[38] See note 27.

[39] I am leaving open the possibility that in special circumstances assuming omniscience may be close enough to correct. The contrast between assuming omniscience in general equilibrium theory and assuming that one is dealing with point particles that interact only by collision in the theory of ideal gases remains. It is not perfectly black and white, but it is stark enough. See note 20.

[40] See 'The methodology of positive economics', pp. 16–19.

[41] Debreu's views were cited before. See also Malinvaud's *Lectures*, p. 242.

[42] *Capital Theory*, p. 301.

[43] This view is connected to the view of economics as employing models which, unlike theories contain no laws. It is a peculiar sort of instrumentalism since it seeks to evade the responsibility of defending the reliability of its tools.

[44] The question of when a theory applies puzzles me. Friedman's view is that theory applies just in case the predictions it makes which are relevant to our concerns are correct (See note 40). I am inclined to the view that a theory applies if and only if its assumptions or antecedents are roughly satisfied. On this latter view we cannot test a theory unless it applied in the test circumstances.

[45] During the presentation of this paper at the Workshop, Professor Edward Green pressed this point lucidly and forcefully.

[46] The sort of 'independence' claimed here is logical, not causal. Useful devices like dated commodities may have been thought up in order to simplify the proofs. Their heuristic value has, however, little connection to the proofs in which they are employed.

[47] *General Competive Analysis*, pp. vi–viii.

[48] I am particularly indebted to Sidney Morgenbesser here.

[49] As Frederick Suppe suggested to me, such curiosity, which motivates the study of general equilibrium in economics, is akin to the physicist's concern with cosmology.

[50] See for example 'Aspects of scientific explanation', pp. 428–430.

[51] 'The winter of our discontent', *Economica* **40** (1973), 324.

[52] If these imaginary economies were without any discernable connection to real economies, the work of the general equilibrium theorists would be of no importance. If the resemblances were quite strong, real economies would approximately satisfy the antecedents of our theories which could then be regarded as straight-forwardly explanatory. For general equilibrium theories to have the sort of value I have been arguing for, we must have something in between.

[53] *Logic*, Bk. VI, Ch. III, sec. 1.

[54] This is the *only* contrast I have drawn between economics and the natural sciences and it is only indirectly related to my claims about the worth of general equilibrium theories.

R. J. WOLFSON

NEW CONSUMER THEORY AND THE RELATIONS BETWEEN GOODS

1. INTRODUCTION

With the exception of the theory of consumer behavior, the entire literature of decision/choice theory is void of concern with the interior structure of outcomes which are being valued and chosen among. Throughout this literature, with that exception, outcomes are treated as atomic wholes. In such treatments the matter of primary interest becomes the process whereby, given certain information, the decision can best be made so as to conform to prescribed criteria which are taken to be the conditions of rational decision. The information which is assumed can be summed up under three headings.

1. The values assigned the outcomes (whether through the intermediation of utility functions or preference orderings).

2. The character of information regarding states of the world which might possibly come about, (ignorance or error[1]), and their probabilities of occurrence (certainty, risk, uncertainty[2]).

3. The character of information regarding the connection between actions taken by the chooser/actor and the ensuing outcome, given the information about states of the world.

But in dealing with consumer behavior the central concern has historically not been with the decision procedure *per se*, but with the impact of 'rational' decision behavior on the demand for goods, under conditions of certainty in most cases, and in the absence of ignorance or error. And insofar as the internal structure of an outcome or market basket is conceived to be just the assembly of goods themselves, this internal structure becomes a matter of concern to the analyst to that degree, and no further. In particular, relationships between goods turn out to be of considerable interest to the economist. These relationships are usually seen to be manifested in the connection between changes in the price of one good and changes in the demand for the other, *cateris paribus*, that is, in terms of the cross-elasticities of demand

J. C. Pitt (ed.), Philosophy in Economics, 33–46.

for the two goods. Moreover, gross substitutability figures in discussions of general economic equilibrium. But in price theory, aside from taking account of cross-elasticities, and investigating connections between these phenomena and the slope and behavior of demand functions, little further is done with them. In particular, no investigation is undertaken of their implications for internal structure of goods.

Substitutability presents few conceptual difficulties, and has been dealt with fairly simply in the theoretical literature. Complementarity is another matter, and while there has been a recognition of its greater complexity, this has been (unfortunately) finessed by allowing it to be conceived of as the opposite of substitutability.

Historically there have been three approaches to these intergood relationships.

The simplest, and probably least useful, is due to Pareto. He looked to the shapes of indifference curves to distinguish complementary (and their 'opposite' – substitute) pairs of goods. Indifference curves which displayed a sharp corner revealed perfect complementarity, and indifference curves which were straight lines displayed perfect substitutability. But this afforded no means for dealing with imperfect cases.

Almost as simple, but somewhat more useful, is the definition based on cross-elasticities. Here, too, complementarity and substitutability are seen as opposites. One good, A, is complementary to (substitutable for) another, B, if associated with a change in the price of B, the cross-elasticity of demand for A is negative (positive). There are two difficulties with this sort of definition. First, it is not necessarily symmetrical. That is, A can be a substitute (complement) for B, while B need not be a (or as good a) substitute (complement) for A. Second, this definition, and the first as well, takes no account of the income effect of price change.

The third definition, due to Hicks[3], is more complex, and difficult to use empirically. But it has the virtue of relying less on market phenomena and more on the individual's preferences for the goods. For Hicks, A is a substitute for B if the marginal rate of substitution of A for money is diminished when B is substituted for money in such a way as to leave the consumer no better off than before. A is complementary with B if the marginal rate of substitution of A for money is increased when B is substituted for money. As Ferguson suggests[4], this is a disequilibrium definition, since it is revealed only if the consumer is not in a state of equilibrium.

This paper is concerned with investigating the character of complementarity and substitutability in the context of two new approaches to the analysis of consumer behavior; one enunciated by Lancaster[5], and one by Becker and others[6].

We will here explore some implications of these two related new formulations of consumer behavior theory – the household production function (HHPF) approach, and the attribute space approach. We will show that the HHPF approach enables us to see *n*-tuples of goods as complements, or as substitutes, as a direct consequence of any cognate production relations they may have in household production processes. But of even greater interest is our finding, upon examining some implications of the attribute space approach, that the relationship of complementarity between goods is not in any fundamental sense an opposite or inverse of the relationship of substitutability between goods. Rather, it is a substitutability relationship between a combination of goods which are complements, and a hypothetical or 'pseudo-' good whose consumption is a goal or target of the consumer.

Before going into this investigation it is profitable to consider why these two approaches to consumer theory are worth the attention of economists and philosophers of social science. What is there about these approaches which makes a recasting of a tried and true apparatus (traditional neoclassical theory of consumer behavior) worth the effort?

There are a number of results which are worthwhile. They are of two sorts. First, there are new capabilities which immediately come to hand as a result of different ways of seeing old relationships. Second, there is a fundamental change in the elements of the theory. The theory is unequivocally more simple. And that has consequences which go beyond aesthetics.

Let us look first at some new capabilities. These approaches, by focussing on the household production commodities which goods make available (i.e., by examining their productivities with respect to commodities), or by focussing on the attributes which goods deliver, make it possible much more easily to analyze processes of market displacement of one good by another. This capability is enhanced by virtue of the better understanding we have gained, as in this investigation, by the relations between goods.

But there is an additional feature of interest especially to philosophers of science, but also increasingly to economists. Both these theories are, in the exact sense in which the term is used by Goodman[7], simpler theories than their antecedents. In traditional consumer theory the number of predicates in the theory is roughly of the order of magnitude of the number of goods which the theory discusses. And, in principle, this number is without limit. In the HHPF theory the number of predicates is roughly of the order of the number of commodities produced in the household. And since our conception of commodities is that they are the basic 'wants' (simple subcategories of 'nutrition', 'clothing', 'shelter', 'leisure', etc.), their number is rather small. Similarly, in the attribute space theory the number of predicates is roughly of the order of the number of relevant attributes of goods. Again, this is a rather small number. So as a result of the adoption of either of these reformulations of consumer theory there is a vast reduction in the number of predicates of the theory. Thus, we have an enormous increase in the simplicity of the theory. As a result, the manifold advantages of simplicity such as greater deductive power and increased consistency of empirical interpretation become available to us.

2. THE HOUSEHOLD PRODUCTION FUNCTION APPROACH

In the household production function approach (HHPF) goods purchased by a household are seen as means to the production in the household of commodities, whose utilities are independent of one another (i.e., are orthogonal in the utility space). It is the commodities which are actually consumed, and demand for goods is a derived demand. That is, the demand for goods is derived from the demand for commodities, and the productivities of the goods in the production of the commodities.

Let Q_j = the quantity of the jth commodity, $j = 1, \ldots, m$,

X_i = the quantity of the ith good, $i = 1, \ldots, n$,

$f_j = (X_1, \ldots, X_n)$ = the production function for the jth commodity,

$U(Q_1, \ldots, Q_m)$ = the utility function of the household,

then

(1) $Q_j = f_j(X_1, \ldots, X_n), j = 1, \ldots, m,$

(2) $U = U(Q_1, \ldots, Q_m) = U(f_1(X_1, \ldots, X_n), \ldots, f_m(X_1, \ldots, X_n))$

(3) $\dfrac{\partial U}{\partial Q_j} = \dfrac{\partial U}{\partial f_j} = U_j = MU_j$ = marginal utility of the jth commodity,

(4) $\dfrac{\partial f_k}{\partial X_i} = f_{ik}$ = marginal productivity of the ith good in production of the kth commodity,

$$(5)\qquad dU = \frac{\partial U}{\partial f_1}\frac{\partial f_1}{\partial X_1}\,dX_1 + \ldots + \frac{\partial U}{\partial f_1}\frac{\partial f_1}{\partial X_n}\,dX_n$$

$$+ \ldots$$

$$+ \frac{\partial U}{\partial f_m}\frac{\partial f_m}{\partial X_1}\,dX_1 + \ldots + \frac{\partial U}{\partial f_m}\frac{\partial f_m}{\partial X_n}\,dX_n,$$

$$(6)\qquad U^{(i)} = \frac{\partial U}{\partial X_i} = \left(\frac{dU}{dX_i}\right)_{dX_r = 0,\, r \neq i} = \frac{\partial U}{\partial f_1}\frac{\partial f_1}{\partial X_i} + \ldots + \frac{\partial U}{\partial f_m}\frac{\partial f_m}{\partial X_i},$$

$$(7)\qquad U^{(i)} = \sum_{j=1}^{m} \frac{\partial U}{\partial f_j}\frac{\partial f_j}{\partial X_i} = \sum_{j=1}^{m} U_j f_{ij} = \text{marginal utility of } i\text{th good.}$$

Suppose that only one commodity is produced in the household. Then any two goods, X_g and X_h, will be substitutes whenever their marginal products both have the same sign (and are nonzero). If their marginal products are zero in any range, and become nonzero (with the same sign) if and only if the use of both is increased, they are complements. That is, these goods are substitutes (complements) if and only if they are substitutes (complements) in the production of the commodity. In this case, then, the price cross-elasticity of demand would reveal degree of substitutability (complementarity) just as it does in the standard analysis of demand.

$$(8)\qquad \eta_{gh} = \frac{\Delta X_g}{X_g} \Big/ \frac{\Delta P_h}{P_h} \quad \text{where } P_h = \text{price of } h\text{th goods,}$$

$$(9)\qquad \eta_{gh} \begin{cases} < 0 \text{ when } X_g \text{ and } X_h \text{ are complements,} \\ = 0 \text{ when } X_g \text{ and } X_h \text{ are unrelated,} \\ > 0 \text{ when } X_g \text{ and } X_h \text{ are substitutes.} \end{cases}$$

If more than one commodity is produced in the household, the same analysis holds, but nothing can be said about substitutability or complementarity between any pair of goods which do not appear together in at least one commodity production function for that household. That is, two goods X_g and X_h will be absolutely unrelated if they are not used together in production of commodities. If they appear together in one and only one household production function, then as before the cross-elasticity is an informative measure of substitutability (complementarity) between two goods. However, if they appear jointly in two or more household production functions, the degree of this inter-good relationship (and, indeed, even the sign) may differ from one HHPF to another. Thus X_g and X_h might be complements in the production of one commodity and substitutes in the production of another. And this would be revealed by the computation of a separate price cross-elasticity of demand for each HHPF in which they both appear.

An overall measure of substitutability or complementarity for a given household or collection of households could be computed, for any pair of goods, simply by computing the overall price cross-elasticity of demand for those two goods. It would, of course, be an average measure, with all the disadvantages of such averages.

Thus far we have reasoned outward from HHPF to price cross-elasticity of demand. Let us see if we can now work in the opposite direction. Suppose that the overall price cross-elasticity of demand is clearly positive (negative). In such a case what we can say about the two goods' involvement in the HHPFs is that in at least some cases they appeared jointly in HHPFs. Moreover, if the overall cross-elasticity is positive (negative), then in at least one HHPF in which they appear together the two goods are substitutes (complements). And it is possible in such a case that they appeared jointly only in that HHPF, or only in HHPFs in which they were substitutes (complements). However, it is also possible that what we have observed is a net effect, that is, that in some of the HHPFs in which they both appeared they are complements, while in others substitutes. If, on the other hand, the overall cross-elasticity is zero, we cannot say that the goods are nowhere related. They may be, but again, local complementarities and local substitutabilities may be cancelling each other out.

Thus, in the HHPF approach to consumer theory the relations between goods are production relations. And measurement of the relationship seems best done, and that not too satisfactorily, by means of price-elasticities of demand.

3. THE ATTRIBUTE SPACES APPROACH

Lancaster presents a plausible view of consumer preference which differs from the standard one.[9] He suggests that the value of goods to consumers is directly associated with the amounts and sorts of characteristics or attributes carried by these goods. In this view, a commodity is seen as a bundle of attributes, as a medium in which are embedded a large (possibly an infinite) number[10] of attributes in various amounts or intensities. Two goods are distinguished from one another by the attributes they carry, and by the amounts or intensities in which the attributes are present in them.

If we assume, for the purposes of this discussion only, that there is a finite class, say exactly n in number, of attributes which are to be found in all members of a set of goods in which we are interested, then any of these goods can be completely characterized by a vector of the amounts or intensities of the corresponding attribute.

In a particular consumption situation some, possibly many, components of the vector (attributes of the good) may be ignored by the consumer. What this implies is that those attributes which are ignored are seen as irrelevant to the good's desirability in that consumption situation. Thus, automotive motor oil surely has a flavor if placed on the tongue. In our notation, then, the vector of attributes which defines motor oil will have some value indicating a flavor, but it is likely to be ignored by consumers of motor oil qua motor oil, although for other goods, such as chocolate Easter eggs, this attribute will be paid significant attention. On the other hand, the motor oil vector's component which registers the coefficient of friction at specified pressures and temperatures will be of considerable interest, while, although there are surely measurable coefficients of friction at those pressures and temperatures for a chocolate Easter egg, who cares? Certainly not the normal consumer of chocolate Easter eggs.

So, any amount of a good can be seen as a vector in an n-dimensional attribute space, where n is the number of 'relevant' attributes. And any bundle of goods can be seen as the vector sum of the several individual goods.

In traditional consumer theory the analysis has been conducted in a commodity space, in which the basis vectors of the space are the commodities themselves. The attribute space is n-dimensional, but the corresponding commodity space is not necessarily of the same dimension. Indeed, it is probably of larger dimension than the attribute space, since the number of distinct commodities (i.e., of non-colinear vectors) which can be generated from a small set of attributes, and which the world does generate, can be very much larger than the number of attributes.

Thus, we have two vector spaces, one of which is of larger dimension than the other, and which are related to each other by a transformation matrix (Lancaster calls it the consumption technology matrix). This matrix transforms the commodity space into the attribute space. It transforms the basis vectors of the commodity space (which are orthogonal in that space) into vectors in the attribute space which are orthogonal in that space only in unusual and trivial circumstances. They would be orthogonal in the attribute space only in the case that the goods were, in no interesting way, related to each other.

So the transformation matrix which carries the commodity space into the attribute space is ordinarily not an orthogonal matrix. It does not preserve orthogonality, its transpose is not its inverse. Moreover, except in the highly improbable case in which the numbers of attributes and commodities are the same, the matrix has no inverse. Thus, while there is a unique transformation carrying the commodity space into the attribute space, there is no unique transformation in the opposite direction. Indeed, even in case the numbers of attributes and commodities are equal, if the transformation matrix is singular then it has no inverse. A singular transformation matrix is one in which there is a linearly dependent set of basis vectors for the commodity space. That is, it would be possible to construct a linear combination (a weighted sum) of some members of that subset, and a linear combination of some other members of that subset such that the sum of the two linear combinations is zero. We are able to construct two vectors which are colinear. When this is possible, what we have done is to construct two bundles of goods in which the relative amounts of attributes are the same. Consequently, the two bundles lie on the same ray through the origin.

3.1. *Substitutability in Attribute Space*

If the two colinear vectors are comprised of single goods, then it seems that we have two goods, each of which presents the same mix of attributes as the other. Now a notion of substitutability emerges from the idea of an attribute space. *Two goods are perfect substitutes if their attribute space representations are linearly dependent, i.e., if their attribute vectors are colinear.*

What if they are not exactly colinear, but are nearly so – a much more likely situation? What can then be said about the degree of substitutability between the two goods?

Clearly, the more nearly colinear the attribute space representations of the two goods are, the more nearly perfect substitutes they are. If they are imperfect substitutes – if their attribute space representations are not colinear, but are less than ninety degrees from each other – their coefficients of substitutability should be less than the maximum (say 1.0) but greater than zero. If they are orthogonal we can say that their coefficient of substitutability should be zero, if they are one hundred and eighty degrees from each other then they are capable of neutralizing each other. In this case we could say that they are perfect negative substitutes, and their coefficient of substitutability is −1.[11] If the two vectors are between ninety and one hundred and eighty degrees apart, then their coefficient of substitutability is negative, but not as small as −1.

There is one vector function which has all these properties, and which, moreover, has some familiar statistical properties whose interpretation conforms to the one just presented. This function is the cosine of the angle between the two vectors. Mathematically it turns out that the cosine is similar to the product-moment correlation coefficient between the attributes of the two commodities.[12] These results are summarized in Figure 1.

Qualitative relation between goods *A* & *B*	Specimen vector diagram (in attribute space)	Smallest angle between vectors (in attribute space)	Desired value of coefficient of substitutability (S)	Value of cosine of smallest angle between vectors
Perfect substitutes	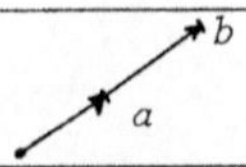	$\theta = 0^0$	$S = 1$	$\cos\theta = 1$
Imperfect substitutes	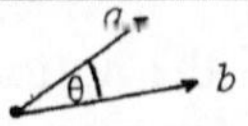	$0^0 < \theta < 90^0$	$0 < S < 1$	$0 < \cos\theta < 1$
Absolute neutrality	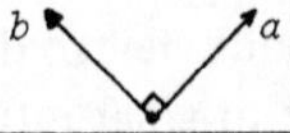	$\theta = 90^0$	$S = 0$	$\cos\theta = 0$
Imperfect negation	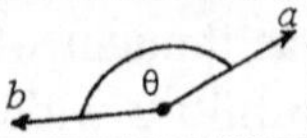	$90^0 < \theta < 180^0$	$-1 < S < 0$	$-1 < \cos\theta < 0$
Perfect negation	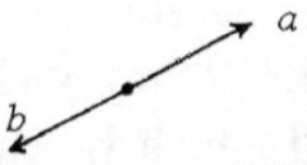	$\theta = 180^0$	$S = -1$	$\cos\theta = -1$

Fig. 1. Substitutability relationship in attribute space.

Strictly speaking, it is not necessary for the complete attribute vectors (i.e., the vectors consisting of the full lists of attributes of the two goods) to be nearly colinear in order for two goods to be good substitutes. What is required is that for each good there be a projection of the attribute vector into the subspace of relevant attributes, which is nearly colinear with the projection of the other good's vector into the same subspace. That is, the two goods should be nearly colinear with respect to the same relevant set of attributes, although if we were to take all attributes into account we might find them far from colinear. Thus, consider the case of two sound reproducing devices. One is a disc-reproducer – an ordinary high fidelity phonograph –

and the other is a cassette tape player. So far as their audio properties are concerned, for purposes of reproducing music they may be nearly identical. The attribute vectors of the two players, projected into a subspace dealing with sound reproducing properties in an ordinary room would be nearly colinear. Hence, in those circumstances they are, for the purposes of sound reproduction, good substitutes. But if we were concerned with operating them in a sports car one more very significant attribute would come into the relevant set – stability in an accelerating framework – and the projections of their attribute vectors in the subspace which is augmented by that one additional dimension would diverge quite significantly.

Thus, the notion of substitutability is quantifiable in terms of attributes, and in order to identify and quantify substitutability we must specify a relevant attribute subspace into which to project the vectors of all attributes of the goods under discussion.

3.2. *Complementarity in Attribute Space*

The foregoing discussion suggests an approach to the notion of complementarity – but not, as heretofore conceived, as the opposite of substitutability. By using the attribute space representation of goods we can develop a notion of complementarity which is consistent with the original intuitive notion of complementarity. Originally, complementarity was thought of as a situation in which two goods, consumed together, yield utilities which complement one another – they perform in a way in which, separately, they cannot, and which is desirable. That is, taken in combination[13] they yield for the consumer a mix of attributes which he desires, and which he cannot get from them separately. In attribute-space terms, they place him near a specific (and desired) attribute ray. They are complements insofar as their combination moves the consumer toward a desired class of vectors. Thus, we see that complementarity, in these terms, requires the designation of a vector as an objective. Two goods are not complementary *in vacuo*. But with respect to a particular target vector which lies somewhere in the attribute space, they may be complements. Their degree of complementarity with respect to the target vector can be measured by the cosine of the angle between the target vector and the vector resulting from the joint consumption (the combination) of the two goods.

One point which emerges from this discussion of complements is that in addition to the two goods whose relationship is being investigated, measurement of their degree of complementarity also requires specification of a third, 'pseudo-' good against which their complementarity is being measured. Moreover, complementarity and substitutability are not, by these definitions, opposites. Indeed, two goods might be reasonably good substitutes (lie fairly close together in an attribute space of relevant basis), and yet, with respect to some target vector, which might or might not be in a different subspace, be complements. For example, consider the case of Bouillabaisse. (Provencal fish stew). A standard recipe[14] for Bouillabaisse calls for the flesh of each of the following seafood:

Red Snapper	Sea Bass	Lobster
Perch	Spanish Mackerel	
Cod	Eel	

With the possible exception of lobsters, any two of the remaining fish might, in other circumstances, be seen by most persons (except for those with idiosyncratic, cultural or religious aversions to some of them) as substitutes of varying degrees of closeness. That is, they might be seen as substitutes for one another as the sole (no pun intended) fish course in a meal. But in the making of Bouillabaisse they are not substitutes. They are all complements, the absence of any one of which would lessen the quality of the Bouillabaisse, i.e., would distance the product from the target.

3.3. *Complementarity, Substitutability and Teleology*

Thus, we see that both complementarity and substitutability, when seen through the glass of attribute-space, must be understood relative to some specification. In the case of substitutability, the specification is in terms of the appropriate or relevant subspace of the full attribute representation of each of the goods. In the case of complementarity, the specification is in terms, not only of the relevant subspace of the full attribute representation of the combination of the two goods, but in addition, of yet another, a target mixture, of the attributes of that subspace. The teleology of the consumer enters both in a clear and explicit way. The teleology of the consumer, in the traditional representations of these inter-goods relations, did enter, but in a murky and undefined way. That is, the teleology of the consumer, as represented by the consumer's 'tastes', his utility function, surely was involved. But the nature of its involvement was quite unclear. This need no longer be the case.

4. CONCLUSIONS

We have explored the impact of two revisions of consumer behavior theory upon substitutability and complementarity, those relations between goods which, in standard neoclassical theory, are taken as given.

The Household Production Function approach to consumer theory substitutes, for the direct impact of goods on the consumption process, the intervention of production processes – the household production processes which convert market goods and household time into commodities which have intrinsic consumption value. Thus, the utility attached to market goods is derived from their productivity of utility in commodities. And substitutability and complementarity among goods derive from substitutability and complementarity among them in production.

On the other hand, the attribute space approach enables us to get a closer look at these relations. Two goods are substitutes for one another insofar as the attribute structures of the relevant subset of their attributes resemble one another. This is not especially surprising, although the notion of the relevant subset of attributes has not been especially dwelt upon heretofore. And it should be understood that this concern brings the consumption context into consideration.

When we turn to complementarity and explore the manner in which it is revealed in the attribute space approach we find some quite interesting results. First, complementarity must be understood as the extent to which two or more goods, in combination, generate a new, desired combination of attributes. Moreover, it becomes apparent that although the price responses of sets of complementary goods are opposite to those of substitutes, complementarity and substitutability are not opposites, as has always been assumed heretofore. And, insofar as complementarity is the degree to which a desired, or targetted state of affairs is achieved, here too we see the entry of the context of consumption.

Finally, in the attribute space representation of utility theory we have been able to construct measures of complementarity and of substitutability which appear to flow naturally from our understanding of the concepts, and which do not depend on market behavior for their computation. Moreover, in contrast to the other measures which have been proposed, these can be computed for sets of any number of goods, and are not vague as to when goods are complements or substitutes.

NOTES

[1] Cf. [17].
[2] Cf. [11, Ch. 13].
[3] Cf. [6, 7].
[4] Cf. [2].
[5] Cf. [8, 9, 10].
[6] Cf. [1, 12, 13].
[7] Cf. [3, 4, 5, pp. 47–54].
[8] Cf. [14].
[9] See also [16].
[10] Although, as is suggested further on, only a small subset of the attributes of a commodity are likely to be of interest to any consumer, or for that matter, to most consumers. In a Ph.D. dissertation in Economics at Syracuse University [15], Tran Huu Dung deals with the relation between attributes and their significance to consumers. Attributes, in his view, are physically accessible on an objective basis. Their significances (i.e., of the attributes, not of their intensities) comprise a vector which transforms the attribute vector and generates the consumer preference function over that commodity.
[11] Although this will be discussed at length later, we note here that such pairs of goods are not complements. However, the notion of complementarity, in general, should not be seen as the opposite, or the inverse, or the negation of substitutability. But some effects of complementarity and substitutability (e.g., cross-elasticity) are opposed.
[12] Consider two vectors $\mathbf{a} = (a_1, \ldots, a_n)$ and $\mathbf{b} = (b_1, \ldots, b_n)$ which form the angle θ

$$\cos\theta = \frac{\mathbf{a}\cdot\mathbf{b}}{|\mathbf{a}||\mathbf{b}|} = \frac{\sum_{i=1}^{n} a_i b_i}{\left(\sum_{j=1}^{n} a_i^2\right)^{1/2}\left(\sum_{i=1}^{n} b_i^2\right)^{1/2}}.$$

[13] 'Combination', a general term, is used here purposefully. It should be understood to mean either of the following:

(a) Simple mixing, so that the attributes of each of the component goods in the mixture emerge unchanged. Consequently, the good resulting from the mixture of the components is defined as the vector sum of the component goods.

(b) Chemical or microphysical combination resulting in changes in some or all of the attributes of the component goods. For example:

Water at 100°C + egg in shell → cooler water + hot egg; rather, we get

Water at 100°C + egg in the shell → cooler water + hot egg whose yolk and white are (to a degree dependent on period of immersion in hot water) partially coagulated

or

$NaHCO_3 + HCl \rightarrow NaCl + H_2O + CO_2$

(sodium bicarbonate plus hydrochloric acid yields sodium chloride plus water plus carbon dioxide, rather than a mixture of the original components).

Thus, it would seem that when such combinations as those described in (b) are made, we must recognize that among the attributes in the component goods are the dispositions to change in certain ways when combined with certain other components. In this case, the dispositions and their triggers yield new attributes, and simple vector addition does not give a description of the resultant good.

14 [18, p. 283].

BIBLIOGRAPHY

[1] Becker, G. S., 'A theory of the allocation of time', *Economic Journal* 75 (1965), 493–517.

[2] Ferguson, C. E., *Microeconomic Theory*, Richard D. Irwin, Inc., rev. ed. 1969.

[3] Goodman, Nelson, 'The test of simplicity', *Science* **CXXVIII** (1958), 150–153.

[4] Goodman, Nelson, 'Recent developments in the theory of simplicity', *Philosophy and Phenomenological Research* **XIX** (1959), 429–446.

[5] Goodman, Nelson, *The Structure of Appearance*, third edition, D. Reidel Publishing Co., Dordrecht-Holland & Boston, Mass., 1977.

[6] Hicks, J. R., *Value and Capital*, Oxford, at the Clarendon Press, 2d. ed. 1946.

[7] Hicks, J. R., *Revision of Demand Theory*, Oxford, at the Clarendon Press, 1956.

[8] Lancaster, K. D., 'A new approach to consumer theory', *Journal of Political Economy* 74 (1966), 132–157.

[9] Lancaster, K. D., 'Change and innovation in the technology of consumption', *American Economic Review (Proceedings)* 52, No. 2 (1966), 14–23.

[10] Lancaster, K. D., *Consumer Demand*, Columbia University Press, 1971.

[11] Luce, R. D. and Raiffa, H., *Games and Decisions*, John Wiley and Sons, Inc., 1957.

[12] Michael, R. T. and Becker, G. S., 'The new theory of consumer behavior', *Swedish Journal of Economics* 75 (1973), 378–396.

[13] Muth, R. F., 'Household production and consumer demand functions', *Econometrica* 34 (1966), 699–708.

[14] Richard S. Rudner, 'An introduction to simplicity', *Philosophy of Science* 28 (1961), 109–119.

[15] Tran Huu Dung, *Toward a Physical Theory of Economic Activities*, Unpublished Doctoral Dissertation in Economics, Syracuse University, 1978, typed, pp. 159.

[16] Wolfson, R. J. and Bower, J. L., 'Studies in attribute analysis: an excursion into the economic analysis of technical change, II. Demand', RAND Corporation D-11442-PR, Santa Monica, California, July 8, 1963, Mimeo, p. 14.

[17] Wolfson, R. J. and Carroll, T. M., 'Ignorance, error and information in the classic theory of decision', *Behavioral Science* 21 (1976), 107–115.

[18] *Gourmet Cookbook*, Gourmet Publishing Co., New York, 1956.

ALEXANDER ROSENBERG

A SKEPTICAL HISTORY OF MICROECONOMIC THEORY

Neo-classical microeconomic theory in its conventional presentation has scarcely changed in nearly a hundred years, although economists have several times shifted their interpretation of its claims, its subject matter and their own intentions in elaborating the theory. In this paper I suggest that the conventional explanations of these shifts belie their real causes, and that the real causes render seriously questionable the viability of the sort of account of economic behavior which is ubiquitous in contemporary economic theory. I do so by offering a sketch of the history of the theory of consumer behavior, and more generally, rational choice theory. This sketch exhibits the shifts in interpretation of the causal variables of the theory, and the changes in the explanatory scope and presumptive subject matter of the theory as largely *ad hoc* qualifications and restrictions that have preserved the theory against a series of failures to empirically substantiate it. I claim the changes to be *ad hoc* in the sense that they purport to be "solutions of all and only those empirical problems which were . . . refuting instances for 'earlier interpretations'."[1] Naturally economists themselves did not advertise these changes as *ad hoc*, but more often than not as changes reflecting improvements in scientific method, especially as reflected in the operationalist imperatives of twentieth century empiricism.

My purpose in offering this piece of interpretative history is not destructive, for I consider the theory under discussion to be the most impressive edifice in social science yet erected. My aim is rather to pose a question: In the light of the facts as I allege them, why do economists continue to commit themselves to this theory and its successors? Surely their aim is not to elaborate a well confirmed systematic nomological account of economic (or other sorts of intentional) action. What then is their aim? What is contemporary neo-classical economics all about?

The marginalist economists built a theory of economic behavior on three distinct general claims: That all agents have complete information about all the alternatives facing them, i.e., they have true beliefs about all facts relevant to their circumstances; they can rank the alternatives these circumstances provide in order of preferability by considering the cardinally measurable utility each will provide the agent; and that their desire to maximize their

J. C. Pitt (ed.), Philosophy in Economics, 47–61.

cardinally measurable utility, combined with their belief that one among the alternative actions available to them will so maximize utility, jointly determine the action they perform. Now the assumption of complete information is a boundary condition, and is plainly unreasonable as a generalization; it was introduced in part only to simplify the construction of theory by enabling the generalization that agents choose those actions which they believe will satisfy their desire to maximize utility to be applied to idealized but nevertheless economically interesting situations. The marginalists did not treat their generalizations about the cardinal utilities of all objects, or events, for individual agents as a simplifying assumption, rather they treated it as an empirical discovery of psychology; similarly, they believed that as a matter of fact humans did act in accordance with the generalization that agents choose that alternative which they believe will satisfy their desire to maximize utility. Of course, they recognized that in the real world, by constrast with their special idealization, perfect information did not prevail, and that agents sometimes made mistakes. But these mistakes reflected mistaken beliefs about the availability or cardinally measurable utility of alternatives, and not the failure of the agent's beliefs and desires to determine his action. The marginalist economists were perfectly clear in their view that in notion of action as utility maximization they had hit upon a formal quantitative version of the general principle that underlies all distinctively human behavior. As Wicksteed put it in *The Common Sense of Political Economics*,

our analysis has shown us that we administer our pecuniary resources on the same principles as those on which we conduct our lives generally. . . . In the course of our investigations we have discovered no special laws of the economic life, but have gained a clearer idea of what that life is.

And further,

Every purchase being a virtual selection and involving a choice between alternatives is made in obedience to impulses and is guided by principles which are equally applicable to other acts of selection and choice. To understand them we must study the psychology of choice . . . . We are constantly weighing apparently heterogeneous objects and selecting between them according to the terms on which we can secure them.[2]

The principle that action maximizes believed available utility is suitably general for the explanation all human action, for it establishes one goal or desire as paramount, and enables us to plausibly explain every other desire cited in a singular causal statement as a desire for some means to this end, or a non-general description of this end, and to assimilate in a general way the factual beliefs about what alternatives are available for attaining this desire.[3]

The economists however, restricted their employment of the principle to an explanation of the demand that agents have for various goods and services. To do this they added a further psychological general claim, that the utility of any item to any agent was a marginally decreasing function of its quantity, and were thus able to deduce that the amount of any good an agent chooses is a decreasing function of the price of that good. But of course because of its generality the principle should be applicable to any human action the strength of whose antecedent desire and content of whose antecedent belief we can determine. This unfortunately proves the Achilles heel of the principle, for we are unable to provide further generalizations which will enable us to measure the cardinal utility of any alternative facing an agent. And without such a further general claim we are unable to apply the principle to actual cases. It is easy to say that what ever the agent chooses is the alternative with the highest cardinal utility, but to so claim is to invoke the very explanatory principle whose applicability is in question in the establishment of its initial conditions of application. While it may seem initially plausible to measure the strength of different desires of a given individual on a scale which like the weight scale has a natural zero, somehow corresponding to indifference to an outcome, it turns out that there are no natural units available to measure cardinal utility. This is just another way of saying there are no generalizations relating distinct amounts of cardinal utility to distinct amounts of something else, which is itself directly or indirectly cardinally measurable. The required generalizations were supposed by (some among) the marginalists to be up to psychology, and in particular psychophysics, to discover, since utility is supposed to bear a relation to felt pleasures and satisfactions. Needless to say this relation is too vague and speculative to be taken very seriously, but the notion that desires qua mental states must differ from each other quantitatively, if they have differing effects on actions chosen is certainly a reasonable one, and offers some justification for the cardinalist view of utilities. The fact that neither introspective, behavioral, cognitive not neurophysiological psychology has yet provided a theory which implies such quantitative differences between differing states of desire,[4] is the real reason why the cardinal notion of utility was surrendered, and why the general law-like claim that agents maximize utility so conceived was surrendered in the course of economics' development.

It has been fashionable for sixty years or more to describe the eclipse of cardinal utility as reflecting its operational meaninglessness, and to describe the advances in economic theory as reflecting a closer attention to empiricist strictures on the introduction of theoretical concepts. Once it was discovered that demand curves could be derived from assumptions about utility weaker than the marginalist's commitment to cardinality, the justification for surrendering the notion came to be largely methodological, and to involve the suggestion that notions of cardinal utility were meaningless because they involved illegitimate reliance on unverifiable introspection. In fact, had economists and psychologists of the period discovered even the most *indirect* nomological connection between differences in the strengths of desires (which we surely know exist), and some naturally measurable differences in behavior or physiology, they would have retained the notion of cardinal utility despite its introspectionist flavor, and operationally independent status. The failure of cardinal utility was simply an empirical one, and not a conceptual or methodological one. It just happens that there are no laws we can discover about cardinal utilities, and so on explanation of actual phenomena can appeal to them. The notion is legitimate and intelligible, but simply reflects a blind alley in the search for the *units* with which to measure one of the causal determinants of behavior. In this respect it was no worse than the pre-Mendelian individuation of hereditary properties; the pre-Mendelian classification schemes resulted in mis-classifying hereditary and non-hereditary traits, just because they hinged on the application of principles of heredity that were empirically false. Just as it took the discovery of the correct theory of heredity to provide the right categorial scheme for hereditary properties, the classificatory employment of a certain conception of utility requires the truth of a particular theory about utilities. If no such theory is correct, the associated classificatory scheme is bound to be empirically unsatisfactory.

Indications that cardinal utility does not characterize a nomologically systematizable notion came in the introspective recognition that on its ordinary interpretation, which made its amounts for any individual a function of the commodities he chooses, this sort of utility is not arithmetically additive, but a function of the availability of other commodities, complements, substitutes, 'superior' and 'inferior' goods, etc.

Systematic sterility of this way of measuring the strength of desires, coupled with its theoretical superfluity for purposes of economics, led to the abandonment of cardinal utility for a weaker notion that required only a measurement of the ordinality of strengths of desires. For it was shown that the downward sloping demand curves that microeconomic theory requires can be derived from hypotheses to the effect that agents can order their preferences from greatest to least, although these desires may bear no intrinsically metrical relation to one another. The appeal to this weaker notion of ordinal utilities was first made by F. Y. Edgeworth and Wilfredo Pareto. One assumes not that individuals attach a numerical quantity of desireability to alternatives, but simply that they rank for preferability each of the alternatives open to them in such a way that they can tell when they are indifferent between varying amounts of any two commodities available to them. The result of this ascription to individuals is the construction of curves of indifference for pairwise comparisons of all available alternatives, and multidimensional surfaces of indifference between amounts of all the commodities available to him. The agent is then assumed to choose that combination of alternative goods, from among those which he believes is available to him, which he most prefers; in the language of indifference curves and budget constrainsts, the combination of commodities uniquely specified by the interaction of the equation determining his budgetary limitations and the indifference surface furthest from the origin among those interesected. The attribution of preference orders and their expression in indifference surfaces represents assumptions considerably weaker than the attribution of cardinal utility, but still general enough to underwrite law-like foundations for explanation of economic behavior, and seemingly extendable to the explanation of actions that are not obviously choices over spatio-temporally discrete goods and services. In particular ordinal utility does not seem to suffer from the nomological and evidential isolation of cardinal utility, for we can measure it independently of the maximization hypothesis in which it functions, or so it seems. To determine an ordinal utility ranking it is not necessary to find a quantitatively varying correlate in physiology or behavior, it seems enough simply to ask the agent to verbally disclose his preferences among alternatives stipulated by the questioner. And even when this is not practicable, it seems reasonable to infer from the counterfactual that the agent could express such a scheme of preferences, that he does have one. Naturally, our 'measurement' of preference structures may not be very systematic nor even produce regularly replicable results, but the attribution of them to individuals on the basis of their actual behavior is as firmly established a feature of our ordinary explanatory apparatus as the assumption that reasons cause human behavior.

Unfortunately, on the assumptions that individuals have such preference orders, and that we can elicit them by constructing a questionnaire, the hypothesis that they always act so as to attain their most preferred objective, subject to their beliefs about availability of alternatives, turns out to be false. It may even turn out to be false in cases where it is reasonable to suppose that the beliefs about alternatives have not changed. In such cases it is often plausible to preserve the maximization hypothesis by hypothesizing so-called 'exogenous' changes in tastes between the time the preference questionnaire was answered and the time the explained or predicted behavior was manifested. But to secure the maximization of ordinal preference hypothesis as something more than an unfalsifiable tautology in the light of these circumstances requires more than just our rough and ready counterfactual belief that had the questionnaire been applied at the moment of actual choice, it would have produced the map of preferences which the actual choice reflects. In short what is required is again a law, a nomological regularity which will connect preference orders and some other variable which can be measured independently of actual choices made as a causal consequence of the preference orders. In the absence of such a regularity to legitimate preference-order as the unit of the causal variable of desire, ordinal utility floats in as much of a systematic void as cardinal utility. It will not do to reply that ordinal utility is know to be a fruitful explanatory notion even in the absence of laws connecting it with other variables measurable independently of desire's causal consequences, because whatever arguments can be constructed for such a conclusion have their companion in arguments for cardinal utility as well. For all such arguments trade on the assumption that anecdotal behavioral evidence for preference rankings can be taken seriously because it reflects the operation of some as yet undiscovered law; but a similar argument will enable us to infer from the existence of another as yet equally unknown law that such evidence also reflects cardinal differences as well as ordinal ones. If the operationalist rejection of cardinal utility is an expression of the failure to discover such laws, then it militates equally as well against ordinal utility measures.

Cardinal utility is often described as having been reflected because it was a notion that relied too heavily on introspection since, because of its unaccessability, introspection is illegitimate a basis for the introduction of scientifically admissible notions. But in fact, the attribution of independent cardinal utilities was rejected because it appeared false on introspective grounds, and no alternative foundation could be provided it. It was after all introspection that convinced economists that commodities' utilities were not independent of one another, and therefore not related in the way required by the simple cardinal notion. Similarly, ordinal conceptions of utility were surrendered with the complaint that they still involved unnecessary or improper appeal to cognitive and conscious states of agents. After all, it was noted, the applications of maximization hypotheses that economics makes surely does not rest on its agents being conscious of and being able to accurately report their preferences; for these assumptions are immaterial to economic problems, and preference-order ascriptions should be interpreted in a way that reflects the irrelevance of these superfluous implications. Thus the surrender of ordinal utility assumptions was described as a species of behaviorist revolution in economics. But the real reason for the economists' surrender of this notion was not the appearance of a new fashionable methodological imperative, but in fact because the reasonable behavioral assumption that ordinal preferences should systematically manifest themselves in behavior like verbal reports turns out to falsify or trivialize the law-like claim that agents choose their most preferred alternative. And the later treatment of desire as a causal variable in the explanation of human action, to which the surrender of ordinal preference eventually led, cannot be described as reflecting the influence of behaviorism in the economists' explanation of human action, but rather as the surrender of any attempt to explain that sort of phenomenon at all. For the account of preference which superceded ordinalist accounts of it, the so-called 'Revealed Preference' theory pre-empts all questions about what desires causally determine individual actions. It does so because, to avoid crediting agents with introspectively available preferences that are publicly inaccessible and yet also undercut our confidence in maximization hypotheses, this theory derives downward sloping demand curves from nothing but the minimal assumption that all agents' actual choices among pairs of alternatives (no matter how causally determined) are always transitive. If from this assumption about the agent's actual behavior, all the economically interesting consequences of traditional marginalist theory follow, then the contemporary economist will eagerly embrace the revealed preference approach; for it absolves him of the duty of finding ways of individuating and measuring the variables which he originally supposed to determine economic (and all other) action, and which he has been unable to isolate over the course of a century's efforts. The price for shirking this task is that the economist is now no longer able to explain his original explanandum phenomenon: individual choice.

Neo-classical economics began with the view, quoted from Wicksteed above, that its subject matter was but as aspect of the general activity of choice, and was to be explained by appeal to laws of a psychological and not a narrowly economic nature. But if the most fundamental assumption economics makes about agents is that their actual behavior manifests a certain transitivity relation, then economics clearly foregoes the resources and the opportunity to explain this behavior, to explain actual choices. If the theory *begins* with actual behavior it is neutral with respect to any theory or competing theories about the causal *determinants* of that behavior, and provides no explanation of it. It is perhaps paradoxical that in the course of attempts to formulate successively more scientifically respectable and empirically warrantable versions of a principle to explain human action in terms of its causal determinants, the end result should be the abject surrender of any pretensions to explain these sorts of events. To parade this turn of events as the harkening of economic theory to the operationalist dicta of modern scientific method and its associated philosophy of science, as many economists have done, is fundamentally to misrepresent the history of a failed research program, by changing the aims of the program after the failure has become apparent. For thirty years after the work of Jevons, Walras, and Menger, cardinal utility was applied to a formalization of what was proposed as a literally true theory of individual choice and its consequence for economic aggregates. After the failure to provide independent specification and measurement for the explanatory variable, it was surrendered in favor or a weaker notion, which held sway for nearly another forty years, though it too suffered from failure, this time of a more complex sort: for the natural interpretation of ordinal preferences turns out to falsify the maximization hypothesis with which it was associated, except when that hypothesis is protected by appeal to indeterminable changes in the exogenous variable of taste.

And so, instead of either surrendering the maximization hypothesis, or continuing to search for an account of preferences that would enable us to apply and confirm the hypothesis, economics surrendered the object of explaining individual economic behavior. It did so in two ways. First, it embraced revealed preference theory as the cornerstone of its theory; and second, the more sophisticated of its theorists began to deny the appropriateness and applicability of the traditional interpretation of its formalism in terms of beliefs and desires of individual agents. They did so by pretending that all along economic theory had no interest in the behavior of the individual agent and that its apparent reference to his beliefs and desires was but a *façon de parler*, a fiction for computational convenience, in the attempt to systematize the real subject matter of economics: markets and economics as a whole. This tradition, which goes back to Marshall, has its most impressive exposition in one of the classic accounts of ordinal utility theory. In *Value and Capital*, Sir John Hicks wrote,

In our discussions of . . . [the law of consumer demand] we have been concerned with the behavior of a single individual. But economics is not in the end much interested in the behavior of single individuals. Its concern is with the behavior of groups. A study of individual demand is only a means to the study of market demand . . . [since] our study of the individual consumer is only a step towards the study of a group of consumers . . . falsifications may be trusted to disappear when the individual account is aggregated.[5]

This attitude, tenable or not, marks a clear break with earlier economists in Hicks' tradition; more importantly, it in no way absolves the economist of the employment of notions like preference, for which he can give no account compatible with the truth of his leading explanatory principle.

On course, experimental psychologists were originally attracted by their own interest in the explanation of choice-behavior to the examination of the economist's explanatory variables, and to his own particular treatment of them. Earliest among them was the distinguished experimentalist L. L. Thurstone, who in the twenties attempted to experimentally infer indifference maps for subjects from behavior, and to predict subsequent indifference maps on their bases.[6] Thurstone's attribution to his (single) experimental subject of an indifference curve between hats and overcoats, hats and shoes, and shoes and overcoats, enabled him to predict an indifference curve for his subject between shoes and overcoats. The method involved assuming that the subject maximizes utility, and that utilities of different commodities are independent (a patent falsehood in the present case), and constructing from preference behavior data utility curves for shoes and overcoats, separately, and then summing them for different combinations of each. Several things are worth noting about this apparently unique experimental attempt to individuate and quantify desires of the sort that the economist and other social scientists require. First, as Thurstone himself noted, the subjects' consistency of choice was remarkable, and he attributed this to the experimenter's careful instruction "to assume a uniform motivational attitude." Secondly, in an attempt to acquire behaviorally based access to paired preferences, the experimenter required appeal to utilities, which are cardinal to the extent that their amounts are assumed to be additive and independent of one another. It is odd that a notion whose behaviorist virtues economists extolled reauired a non-behaviorist notion for its isolation and description. This being the case, the special methodological merits of ordinal utility turn out to be illusory. Finally, the assumptions of utility maximization and of utility independence may be claimed to rule out Thurstone's method of measuring preferences, as one which will enable us to assess the merits of the economist's hypothesis of utility maximization; for it clearly begs this very question. I say that this charge may be laid, but not that it must be, or will be decisive, because all scientific measurement involves a circle of notions and assumptions, and the real questions about their adequacy are whether the circles are large enough to provide their elements with explanatory power.

But the main limitation on Thurstone's experimental technique for constructing indifference curves, aside from the difficulty of generalizing them from paired comparisions, to *n*-tuples of the dimensions agents actually face, and aside from the fact that in the absence of careful instructions, preference behavior turns out to be inconsistent and insensitive, is that it enables us at best only to predict the subject's indifference curve for choices between combinations of commodities already offered in conditions of certainty. That is, it enables us to predict no actual choices among these commodities in conditions of uncertainty. It is surely no objection to Thurstone's method that his experiment was closed to the effects of uncertainty in the agent's beliefs, for it is only reasonable in attempting to find a behavioral measure for one psychological variable that it should be allowed to vary in circumstances from which another co-determinant of the same behavior is excluded. The problem is, when we come to measure the second variable, belief, whether we can hold desire constant in the same way. If we cannot our confidence in the ability of these two variables to systematically and precisely explain the consequences we accord them will be seriously undermined.

Closing the explanandum-phenomena to the effects of variations in belief has been a central characteristic of economics since the time of the marginalists. The ubiquitous assumption that all agents have perfect information about available alternatives is nothing less than the attempt to insulate the explanation of consumer behavior against the influences of doxastic states of varying degrees of strength, as well as different contents, and with varying degrees of influence on behavior. And just as such insulation was reasonable for psychologists like Thurstone, it was equally legitimate during the period in which economists were attempting to clarify the identifying marks of variations in the strength of preferences in behavior. Naturally, it is reasonable to expect that once ways of measuring each of these distinct causal forces are discovered, we will be able to bring them together in tests of the claim that they are jointly sufficient for the nomological determination of human action. In effect, our expectation is identical to the expectation, eventually fulfilled, that measurements of pressure and temperature of a gas, independent of each other, and of volume of the gas, could both be employed to establish that volume is a nomological function of pressure and temperature.

Although economists never did solve the problem of measuring strength of desire independent of behavior, and psychologists likewise made little progress on the problem, both recognized the importance of providing measures of strength and content of beliefs, for both eventually began to focus on the problems of decision under conditions of imperfect information. Of course, neither economists nor psychologists expressed the problem of quantifying the causal variable of belief in this way, rather, they set themselves the problem of waiving the assumption of perfect information, and explaining human action under conditions of uncertainty and risk. 'Risk' is generally used to describe those cases in which our beliefs about alternatives available can be given a probabilistic measure that reflects our degree of confidence that these alternatives will actually be available. 'Uncertainty' is often used to describe situations in which our knowledge of the availability of alternatives does not even extend to the attachment of probabilities; these are situations in which we have almost no knowledge about availability of alternatives. The most important theory of decision under conditions of risk is of course that of John von Neumann and Oscar Morgenstern.[7] In effect, they showed that it is possible to construct a utility measure, which, together with the maximization of utility hypothesis, determined agents' actions in the face of alternatives to which they can attach probabilities. Determinate choices under conditions of assessable risk follow from five axioms about agents' preferences among alternatives available with certainty, and/or with varying degrees of probability as well as combinations of alternatives available with certainty or probability. Now, while this model represents a formulation of the traditional economist's maximization hypothesis that enables us to apply it to the explanation of action in the absence of certainty of belief about outcomes, and which is therefore much more relevant to the explanation of actual behavior than the original hypothesis, it not only does not avoid the specification problem for preferences that bedeviled the marginalists, it also generates a specification problem for the probabilistic beliefs it accords to agents. In the absence of such a specification, actual predictions of choice under conditions of risk cannot be made, because we cannot measure the probabilities which agents assign to alternatives, which, in their beliefs, may be available. Of course we can infer their beliefs from their actual choices, on the assumption that their preferences satisfy the five axioms of the von Neumann–Morgenstern theory, and that they maximize the (so called 'expected') utility which that theory accords agents, but such an inference will not allow the independent specification of probability assignments that we require. What is worse, the axioms seem independently to be falsified by the apparent fact that some agents derive pleasure from the sheer opportunity to take risks, even at losing odds, and others avoid risk at any cost. Such agents seem to violate the von Neumann–Morgenstern axioms because they are never indifferent in their preferences between a given alternative with certainty and a combination of more and less preferred alternatives with given probabilities. The axioms 'seem' to be falsified by apparent risk avoidance and risk attraction, but we really do not know whether they are or not, because we have no independent measure of either preferences or subjective assessment of probabilities which we require in order to apply and test the von Neumann–Morgenstern account.

Just as the economist's problem of specifying preferences was reflected in experimental work of psychologists, similarly, the notion of subjective probability as a measure of strength of belief also attracted the attention of psychologists. It has however been an invariable feature of all the sustained accounts of subjective probability attributions that they all rest on a prior attribution of some preferences order or utility structure to their subjects. That is, each involves the employment of actual behavior as a measure of strength of belief, but only on the assumption that the experimental subject has, for example, a specified utility scale for money, or for some other reward or reinforcement. A number of independent experiments performed in the late forties and fifties[8] showed considerable agreement on the question to how subjective estimates of probability vary with objective, actual probabilities, but each involved the assumption that money pay-offs determine a preference structure, and so none provide a measure of strength of belief independent of assumptions about the existence and causal force of varying strengths of desire. The potentially inextricable connection between belief and desire is reflected in work by philosophers and mathematicians on the foundations and interpretation of probability and statistical theory. Thus, for instance Frank Ramsey's influential paper 'Truth and Probability'[9] involves explicit appeal to utilities in the measurement of probabilities; and L. J. Savage's *Foundations of Statistics*[10] propounds a measure of probability of all events on the assumption, among others, that all actions can be ordinally ranked for preference. Of course there seems to be an alternative in the determination of subjective probabilities to complex inferences from choice behavior that presupposes preferences: just ask the subjects to communicate their strength of belief by making verbal estimates of probabilities. This method is akin to the notion that we can measure preference strength independent of behavior simply by asking the subject to verbally rank available alternatives. Experiments have been performed to test hypotheses about the relationship between subject measures of probability and actual objective probabilities, simply by asking subjects to estimate the frequencies which are known independently by the experimenter. The general conclusion has been that subjective measures of probability are roughly a linear function of objective probabilities, although not identical to them, and that they vary over individuals.[11] The trouble with such specifications of strength of strength of belief is identical to the problems associated with verbal reporting as a measure of strength of preference. If accepted they show maximization of utility or preference hypotheses to be false. On the other hand, the applicability and descriptive accuracy of a maximization hypothesis is not held by economists or psychologists to rest on agent's abilities to be continually conscious of or able to accurately report the estimates of probability which govern their actions. Thus, direct questioning is rejected as a source of identification for one of the causal determinants of behavior on the grounds that it makes methodologically illegitimate appeal to phenomena that are only introspectively available, and cannot be publically confirmed.

Economists seem to be faced with the following difficulty. Their employment of any law-like claim to the effect that actions reflect the maximization of preferences, subject to beliefs about alternatives, requires a specification of the beliefs and desires in question. And yet none of the specifications of desires that trades on our common-sense understanding of this notion seems to confirm the maximization hypothesis, and what is worse, no specification of strength of belief seems available in which that notion is isolated in the required way from the notion of strength of desire. Accordingly, the maximization principle can be preserved only on the condition that it is not actually applicable to the practical prediction and control of any particular human action, for we do not have the resources to establish the initial conditions for such predictions. In the absence of the required auxiliary hypotheses economists found a number of ways to preserve the employment of maximization hypotheses. One of these ways involved seeking the protection of behaviorist and operationalist methodological dicta to exclude interpretations and specifications of belief and desire that disconfirm the hypothesis. Sometimes this tactic has been joined to the proclamation that their theory is not to be understood as dealing with the actual behavior of individual agents. The trouble with such proclamations are at least two fold. The first is that it is rejected by many economists and belied in their actual work; so much so that it seems more a rationalization for their failure to find the required specifications, than an expression of the aims and claims of economics. Second, and more important, in searching for a general law relating action with reasons, economic science has been doing duty for all the other sciences that trade in these same causal variables, and yet do not explicitly formulate the law which a commitment to empiricism requires underlie them. Insofar as the other social sciences actually do purport to provide explanations of individual human action by appeal to beliefs and desires, they can hardly adopt the economist's disdain for individual human action as a fitting subject of study.

Microeconomic theory reaches its apogee with the development of its theory of market clearing general equilibrium. The fascination that general equilibrium theory exercises for contemporary economists belies much of the interpretative shifting that is recorded in these pages. And its importance for economists reflects as much as these shifts the question of what it is exactly that economists are up to. For theirs is clearly not the attempt to develop an empirically controlled theory of economic behavior; it seems much more like an exercise in mapping out the formal consequences of a set of topologically expressible axioms whose evidential basis and explanatory power is almost completely irrelevant to the economists' interest in them. What then are the aims of this subject so little changed in over a hundred years?

Syracuse University

NOTES

[1] I employ here Lawdan's characterization. Cf. *Progress and Its Problems* (California, 1978), p. 114.

[2] P. H. Wicksteed, *Common Sense of Political Economy* (London, 1910), pp. 126, 13.

[3] Of course neither utility nor the maximization of utility is the intentional object of desire for anyone but a trained neo-classical economist. But, according to the theory, every intentional object of desire is desired because it is a (perhaps nonintentional) means to this end of maximizing utility. In effect, the theory explains intentional purposive behavior as a species of pentially nonintentional teleological processes. The logical and semantic niceties of this matter are the purview of the philosophy of mind and are beyond present scope. Moreover, they leave untouched the claims of this paper.

[4] Of course many hypotheses and speculations about the motivational determinants of human behavior have come and gone in the varying branches of psychology. And this claim is not meant to deny their existence. The fact is none of these theories have met with general acceptance, and none has deserved to do so. For none provides hypotheses relating cognitive states to measurable variables in such a way as to increase the power of explanation and of prediction for human actions beyond the bounds of commonsense's abilities to do so.

[5] Sir John Hicks, *Value and Capital* (Oxford, 1913), p. 11.

[6] L. L. Thurstone, 'The indifference function', *Journal of Social Psychology* 2 (1931), 139–167.

[7] *Theory of Games and Economic Behavior* (Princeton, 1944).
[8] These experiments are reported and discussed in Ward Edwards, 'Theory of decision making', *Psychological Bulletin* 51 (1954), 397ff.
[9] In *Foundations of Mathematics and Other Logical Essays* (New York, 1931).
[10] (New York, 1954).
[11] Experiments reported in Ward Edwards, 'Behavioral decision theory', *Annual Review of Psychology* (1964), 479.

STEVEN STRASNICK

NEO-UTILITARIAN ETHICS AND THE ORDINAL REPRESENTATION ASSUMPTION

1

Utilitarianism is a very resilient theory. Once the darling of both philosophers and economists, it lost favor in the first half of this century as methodological developments in both fields undercut its conceptual standing. From the economist's point of view, the strictly ordinal judgements of the neo-classical school with their indifference curves seemed all that were required to analyze the rational behavior of individuals and groups.[1] The concept of cardinally significant individual utilities was viewed as just so much excess baggage for Occam's famous razor to slice away, since in the real world all we can ever observe is choices and not the intensity of choices. Similar trends in philosophy caused the logical positivists to discard as meaningless any concept that could not be empirically verified. But in the fifties, riding the crest of a new economic theory of rationality, utilitarianism staged its comeback. In their famous 1944 work, *Theory of Games and Economic Behavior*, von Neumann and Morgenstern showed that individual choices could in fact be used to define a cardinal numerical utility, if these choices were among uncertain prospects and satisfied weak consistency postulates of rationality.[2] Building upon their work, the economist Harsanyi showed that these postulates would, when coupled with a weak ethical postulate, logically entail that social welfare was an additive function of individual utilities.[3] Similar results for sure prospects were obtained by the economist Fleming, using somewhat stronger ethical postulates.[4]

Confronted by the powerful theorems of the neo-utilitarians, a philosopher who shares the common view that utilitarianism is insensitive to the demands of distributive equity would be faced with somewhat of a dilemma. A sophisticated account of this view has recently been developed by Rawls in his *A Theory of Justice*, which argues, among other things, for the difference principle, a maximin-like principle of social justice which tells us to maximize the prospects of the worst-off class in society, as opposed to the sum of individual utilities.[5] But Rawls' view and all those similar to it must be abandoned if the results of Fleming and Harsanyi are acceptd. Could it be that, intuitions about justice to the contrary, there is something about the

J. C. Pitt (ed.), Philosophy in Economics, 63–92.

logical structure of rational decision that compels any minimumly acceptable ethical system to be utilitarian?

To escape the above conclusion, philosophers are beginning to challenge the neo-utilitarian account of rationality, focusing especially, in the von Neumann–Morgenstern framework, at least, on that postulate which seems to be decisive for establishing the cardinality of utility over its mere ordinality, namely, the so-called sure-thing principle. While there may in fact be problems with this principle, I shall argue that the real culprit in the neo-utilitarian account is an assumption that was not even added by Fleming or Harsanyi to the ordinalist account, but was instead already present. For the ordinalists assumed that if individual or social choices were consistent enough to define an ordering among prospects, then these choices should be representable by an ordinal utility function which assigns a real number to each prospect. As we shall see, this innocent looking assumption about the possibility of numerical representation, which we will call the *ordinal representation assumption*, conceals an unreasonably restrictive assumption about the one dimensional nature of rational choice, whether individual or social, an assumption which inexorably leads to the neo-utilitarian position. If a more general account of rationality is adopted, then it will be the case, as the neo-utilitarians argue, that the logical structure of rational decision will require any acceptable ethical system to have a certain form. But this form will be one more akin to Rawls' difference principle than to classical utilitarian ethics.

2

In order to see the significance of the ordinal representation assumption in the neo-utilitarian argument, let us consider Fleming's analysis in his 1952 paper, 'A Cardinal Concept of Welfare'. Besides being noteworthy for its role in the revival of pre-Pareto utilitarian thought, Fleming's paper is also significant because it represents one of the first applications of the axiomatic method to moral theory. For it was Fleming's intention to prove that any ethical system which conformed to certain weak conditions provided, through the process of ethical evaluation itself, a manner of measuring individual well-being with the property that the social welfare associated with a state would be some additive function of these individual well-beings. In the tradition of Bergson, Fleming viewed his task as determining how the evaluation of social welfare (or desirability, as he called it) should reflect information about individual welfare.[6] As Bergson defined it, social welfare was a fairly open-ended concept, and Fleming sought to give it more content by imposing conditions on the manner in which an ethical system made judgements of social welfare. Essentially, Fleming imposed two different kinds of constraints. The first set concerned a series of consistency relationships that judgements of social welfare should manifest vis-à-vis each other, while the second set concerned the relationship of individual well-being and social welfare. In stating Fleming's postulates, we shall use the more familiar language of preference in place of Fleming's terminology of desirability and well-being.

Suppose 'xPy' represents the claim that situation x is preferred to situation y from a social standpoint and 'xIy' the claim of social indifference. Then we can state Fleming's first set of postulates as follows:

POSTULATE A. If xPy, then not yPx.

POSTULATE B. If xPy and yPz, then xPz.

POSTULATE C. If xIy and yIz, then xIz.

Fleming's first postulate establishes the asymmetry of social preference, while the second and third establish the transitivity of social preference and indifference respectively. Fleming obviously thought that these postulates were sufficient to insure that social preference would establish an ordering among all possible situations in terms of their relative desirability, or what he called a "linear order of precedence." Strictly speaking, though, his postulates establish only a partial ordering. For a complete ordering defined for all situations, Fleming needs a completeness requirement, such as: for any x and y, either xPy or yPx or xIy. Otherwise there may exist situations which were incomparable from the social standpoint.[7]

Fleming's second set of postulates may be viewed as representing certain ethical requirements about the relationship of individual and social preferences. Let 'xP_iy' represent individual i's preference for x over y and 'xI_iy' his indifference between the two. Then Fleming's next postulate will take the form of the well-known strong Pareto principle.

POSTULATE D. (a) If for all i, xI_iy, then xIy; (b) If for all $i \neq j$, xI_iy and xP_jy, then xPy.

Fleming's final postulate strengthens the strong Pareto rule by requiring that the social preference between two situations should depend only on the preferences of individuals who are not indifferent between them. We can state this requirement in the following manner, without altering Fleming's intent:

POSTULATE E. Suppose we can partition the set S of all individual preferences among x and y into two mutually exclusive and jointly exhaustive non-empty subsets S_1 and S_2 such that for all i in S_1, $x I_i y$, and for no i in S_2, $x I_i y$. Then the social preference defined for the set S must be identical to that defined for S_2.

This postulate has been called the principle of the elimination of indifferent individuals.

Given these postulates, Fleming claims that he can show that

> for any ethical system which is compatible with postulates A to E, it is possible so to measure welfare and individual-momentary well-beings that welfare is the sum of well-beings, and that the measurements so arrived at are unique save for proportionate scale (p. 255).

To establish this claim, Fleming goes on to show how one may construct functions measuring individual well-being and social welfare which stand in the required additive relationship. While Fleming does construct the required functions, his construction does not depend solely on postulates A to E. Instead Fleming must make use of that hidden representational assumption I referred to earlier, namely, the ordinal representation assumption. For without this assumption, it is easy to show that there exists an ethical system which is not related to individual well-being in the manner Fleming's result requires.

3

The essence of Fleming's claim is that any ethical system which defines an ordering among all possible social situations and satisfies postulates D and E will require measures of individual well-being that are cardinally significant. If we assume that individual well-being can only be compared in an ordinally significant manner, then it should not be possible for us to define an ethical system that satisfies postulates D and E and establishes an ordering over all social situations. If we can show the existence of such a system, then Fleming's construction will be invalid.

Accordingly, let us assume that at best we can make only ordinally significant judgements concerning the relative well-being of individuals in the various social situations. Thus, for any two individuals i and j in any two situations x and y, if x_i represents the relative level of well-being of i in x, and y_j the relative level of well-being of j in y, we will be able to make only one of the following judgements concerning their relative well-beings: either

$x_i > y_j$, $x_i = y_j$, or $x_i < y_j$. On the basis of these strictly ordinal judgements, which have been called extended sympathy judgements, we will be able to define an extended ordering of the relative level of well-being of all individuals in all situations.[8] Let $\tilde{R}$ represent this ordering and R the ordering of all situations determined by social preference. Then if Fleming is right, there should not exist a function f such that $f(\tilde{R}) = R$ which satisfies postulates D and E.[9]

But consider the following ethical principle, which we will call the lexical maximin principle, or leximin, for short. For every situation x, let $R(x)$ represent the ordering of the relative well-beings of all n individuals in x on the basis of the extended ordering $\tilde{R}$. Assign an integer ranking to each individual i on the basis of his position in $R(x)$, with the worst-off individual assigned the ranking 1, and ties between equally well-off individuals broken arbitrarily. Then let $r(x)$ represent the r-th worst-off position in situation x. We may now state the following principle:

LEXIMIN PRINCIPLE. For any x, y, if there is a k: $1 < k < n$ such that $x_{k(x)} > y_{k(y)}$, and for all $r < k$: $x_{r(x)} = y_{r(y)}$, then $x\mathrm{P}y$. If, on the other hand, for all r : $1 < r < n$, $x_{r(x)} = y_{r(y)}$, then $x\mathrm{I}y$.[10]

This principle requires that situation x will be socially preferred to y if the worst-off individual in x is better off than the worst-off individual in y. In case of ties, the social preference must look to the relative positions of the next worst-off, and so on. Since the leximin will lexicographically maximize the well-being of the worst-off individual, it is in fact a lexical version of that principle mentioned at the beginning of the paper, i.e., Rawls' difference principle.

Clearly, given $\tilde{R}$, the leximin will define an ordering R, so that Fleming's postulates A through C are satisfied. Moreover, if we make the natural assumption that $x\mathrm{P}_i y$ iff $x_i > y_i$, $x\mathrm{I}_i y$ iff $x_i = y_i$, and $y\mathrm{P}_i x$ iff $y_i > x_i$, we can see that the leximin will satisfy postulates D and E as well. Part (a) of postulate D, which requires that $x\mathrm{I}y$ if for all i, $x\mathrm{I}_i y$, will be satisfied automatically, since it will be the case that for all r: $1 < r < n$, $x_{r(x)} = y_{r(y)}$. Similarly, part (b) will be satisfied because if for all $i \neq j$, $x\mathrm{I}_i y$ and $x\mathrm{P}_j y$, if we assume $y_j = y_{k(y)}$, it will be the case that $x_{k(x)} > y_{k(y)}$ and for all $r < k$, $x_{r(x)} = y_{r(y)}$. Finally, let us consider the nature of the social preference for the partition S_2 of non-indifferent individuals, as defined by postulate E, and let $r'(x)$ be the r-th worst-off individual in S_2. If we assume a social preference of $x\mathrm{P}y$ for S_2, then there will exist a k' such that $x_{k'(x)} > y_{k'(y)}$ and for all $r' < k'$, $x_{r'(x)} = y_{r'(y)}$. Of course, if there are no individuals i and j in S_2 such that

xP_iy and yP_jx and $y_i = x_j$, for example, then it will be the case that $k' = 1$. Now let us add to S_2 the set of indifferent individuals S_1 and consider the nature of the social preference for the complete set S, using the usual notation $r(x)$. If among the indifferent individuals there is no individual h such that $y_h < y_{k'(y)}$, then $k' = k$ and clearly xPy. On the other hand, if there are m such individuals, then $k = k' + m$, though it will still be the case that $x_{k(x)} > y_{k(y)}$ and for all $r < k$, $x_{r(x)} = y_{r(y)}$, and thus xPy. We see then that the leximin satisfies postulate E as well.

4

Since we have defined a function f which translates purely ordinal judgements concerning individual well-being, $\tilde{R}$, into a social ordering R, and which at the same time satisfies postulates D and E, it is clear that Fleming must add an additional assumption to his postulates if his argument is to be successful. In fact, if one examines Fleming's paper, it is apparent that Fleming thought this additional postulate followed logically from his postulates A through C, which required that social preference define an ordering over all possible situations. Consider the following passage, where Fleming discusses the implication of postulates A through C being satisfied by the property 'blankness':

> If all of these requirements are met, we may speak of blankness as an ordinal magnitude, different degrees of which may be distinguished, occupying different positions in a linear order. Numbers can then be assigned to the degrees of blankness possessed respectively by different objects, in such a way that the same number will be assigned to objects of the same degree of blankness, and that the blanker of two objects will always be assigned the higher number. Such numbers may be regarded as particular values corresponding to particular degrees of blankness, of a continuous function, or indicator providing a cardinal representation of the property 'blankness' (p. 246).

In this passage, Fleming commits an error that was fairly common during the time he was writing, since he assumes that the ordinal representation assumption follows without argument. For example, when Harsanyi discusses Fleming's postulates A through C in his 1955 paper, he makes precisely the same mistake:

> These three postulates are meant to insure that 'social preference' establishes a complete ordering among the possible social situatiors, from which the existence of a social welfare function (at least of an ordinal type) at once follows (p. 268).

Fleming made this mistake, of course, for it was his aim to define a measure of the concept 'social welfare', in the spirit of Bergson's social welfare function.

For Fleming, the amount of social welfare a situation possessed was equivalent to the degree of its desirability, a simple one-dimensional property that each situation possessed to some degree. The fact that one situation was socially preferred to another was assumed to represent the fact that the one had more 'desirability' or 'social welfare' than the other. Since social preference was determined on the basis of this relative 'desirability', it seemed reasonable that there should exist a social welfare function which assigned a number to represent the degree of this desirability.

The trap Fleming has fallen into is obviously a very natural one, and it is hard to see what is wrong with the ordinal representation assumption in the context of the concept of evaluation that it models. After all, doesn't the pure concept of morality look like a prime candidate for representation by something akin to the social welfare function? When we say that one situation is morally preferable to another, don't we just mean that the one has more moral 'desirability' that the other? Perhaps there are occasions when we do make judgements on the basis of our evaluation of relative moral 'desirability', but it is also clear that sometimes our moral judgements have an entirely different basis, one that is not susceptible to representation by something like the social welfare function. The type of basis I have in mind is sometimes called a *deontological* one by philosophers and concerns the priority of claims.[11]

For, suppose there exist situations where the interests of individuals are in conflict and where the correct moral resolution of the conflict depends on a moral weighting of the competing claims. If one individual's claim is judged to be of greater priority than the claims of others, on the basis of this individual's greater desert or merit or because other individuals are bound through past obligations, for example, then the fact of this greater priority will be sufficient to determine the correct moral outcome, irrespective of whether there exists some measure of the relative 'desirability' of this outcome or not. If this view of moral evaluation is ever appropriate, then the model of morality based on the social welfare function must be supplemented by another kind of model, one that is appropriate for representing a priority weighting of different claims.

To see why another model would be required for representing priority claims, consider the following simplified case of moral judgement. Suppose a group of individuals are competing for some scarce good and that we want to rank these individuals in terms of their relative priority. Each individual has two types of claims that he can make against this good, C^1 and C^2, with C_i^1, C_i^2 representing the strength of individual i's two claims and taking values between 0 and 1. Finally, suppose claims of the first type are viewed as taking precedence over claims of the second type, so that individual i will always be ranked higher in priority than individual j whenever $C_i^1 > C_j^1$. If $C_i^1 = C_j^1$, however, individual i will be ranked higher just in case $C_i^2 > C_j^2$. Clearly, by using these priority rules we will be able to define a complete ranking of individual priority for all possible values of C_i^1 and C_i^2. *But while this is obviously* a reasonable method of moral judgement, given the priority of C^1 claims over C^2 claims, it is one that is not representable by a function which assigns to each individual a number s representing the strength of his respective claims and ranks one individual higher than another just in case his s-value is greater. Debreu makes this clear in a note from his 1954 paper, 'Representation of a Preference Ordering by a Numerical Function':[12]

> Consider the lexicographic ordering of the plane: a point of coordinates (a', b') is better than the point (a, b) if '$a' > a$' or if '$a' = a$ and $b' > b$'. Suppose that there exists a real order-preserving function $\alpha(a, b)$. Take two fixed numbers $\alpha_1(a) = \alpha(a, b_1)$ and $\alpha_2(a, b_2)$. To two different numbers a, a' correspond two disjoint intervals $(\alpha_1(a), \alpha_2(a))$ and $(\alpha_1(a'), \alpha_2(a'))$. One obtains therefore a one-to-one correspondence between the set of real numbers (*non-countable*) and a set of *non-degenerate disjoint* intervals (*countable*) (p. 164).

Thus, if a moral system is to take into account the fact that some kinds of claims have priority over others along the lines of the simple case sketched above, it will not be representable by a function that maps the moral 'desirability' of situations onto the reals.

Given the above example, it should now be clear why the assumption of numerical representability for the ordering defined by social preference is so critical to Fleming's result. For the lexical version of Rawls' difference principle is in fact a kind of claims-based moral system. If we assume that a real-valued measure of individual well-being in the various situations exists, then this measure also becomes a measure of the strength of the individual's claim for satisfaction. The ranking $R(x)$ that we introduced earlier becomes a ranking of the priority of individual claims for the realization of situation x. And since, from the point of view of Rawls' difference principle, stronger claims must be satisfied first, one situation will be socially preferred to another if the strongest claim for the first situation is stronger than the strongest claim for the second. In case of ties, the next strongest claims are considered, and so on. Because the lexical difference principle has this claims-based structure, it will not be representable by a real-valued social welfare function.[13] Thus, Fleming's seemingly innocuous assumption about numerical representability has the consequence of ruling out the lexical difference principle as an acceptable moral system. And it is this fact which then allows him to conclude that any minimumly acceptable ethical system will be some additive function of cardinally significant measures of individual well-being.

5

Given the role of the ordinal representational assumption in Fleming's argument, we can now reevaluate the significance of Fleming's result. Fleming does not in fact establish that any ethical system that satisfies his postulates A through E must be an additive function of cardinally significant measures of individual well-being. Rather, he shows that any ethical system that is representable by something like the social welfare function and satisfies these postulates must have this form. In a sense, Fleming's conclusion about the relationship between social welfare and individual well-being is not so much a consequence of his postulates as it is a consequence of an assumption about the logical structure of moral judgement that was itself embedded in the ordinal representational assumption. If we can make this structural assumption explicit, then perhaps we will be able to better understand why Fleming's result produces the relationship between social welfare and individual well-being that it does.

Accordingly, under the assumption that one situation will be socially preferred to another just in case the former has more moral desirability associated with it than does the latter, we shall define what we will call a moral desirability function, or an MDF. Let $d_i(x, y)$ represent the degree to which individual i's preference contributes, either positively or negatively, to the judgement that situation x has more moral desirability than situation y. Then, the MDF will have the following structure:

MORAL DESIRABILITY FUNCTION. For any alternatives x, y, the social preference will be

(a) xPy iff $d_1(x, y) + d_2(x, y) + \ldots + d_n(x, y) > 0$;

(b) xIy iff $d_1(x, y) + d_2(x, y) + \ldots + d_n(x, y) = 0$;

(c) yPx iff $d_1(x, y) + d_2(x, y) + \ldots + d_n(x, y) < 0$.

Given the manner in which $d_i(x, y)$ has been defined, of course, the form we have chosen for the MDF is an entirely natural one. Each individual's $d_i(x, y)$ represents the extent to which his preference contributes to the judgement that situation x has more desirability than y, and if over all, the balance of such contributions favors x over y, then xPy should be the social preference. What is not so clear, of course, is why moral judgement should have this kind of structure, or in other works, why a function $d_i(x, y)$ should exist, as opposed to a function, for example, which represents the strength of i's *claim* for x over y. Let us call $d_i(x, y)$ the individual's *aggregative index* and the assumption that $d_i(x, y)$ exists the *aggregative existence assumption.*

Given the aggregative index and thus the MDF, let us now investigate some of its consequences by placing some weak ethical constraints on $d_i(x, y)$. First, suppose that there exist measures of individual well-being x_i, y_i and that xP_iy iff $x_i > y_i$, etc., as before and that $d_i(x, y) = d_i(x_i, y_i)$. And assume that $d_i(x_i, y_i)$ will be the same function for all individuals i, so that we may write $d(x_i, y_i)$. We shall require $d(x_i, y_i)$ to satisfy the following two conditions:

PERMUTATION CONDITION. For all x, y and i, j, $d(x_i, y_i) + d(x_j, y_j) = d(x_j, y_i) + d(x_i, y_j)$.

Since the equality of the two sums will entail that the social preference will be the same for the cases represented by the two sides of the equation, this condition is a kind of impartiality condition and requires that the social preference will remain invariant under permutations of well-beings among individuals. The second condition will represent the force of the Strong Pareto condition in the context of the MDF:

MONOTONICITY CONDITION. For all x, y and all i,

(a) if $x_i + y_i$, then $d(x_i, y_i) = 0$;

(b) let c be some positive real-valued number. Then $d(x_i, + c, y_i) > d(x_i, y_i)$ and $d(x_i, y_i) > d(x_i, y_i + c)$.

To the weak requirement that indifferent individuals not contribute to judgements of desirability, this condition adds the further requirement that the function d respond in a positive manner to increases in individual well-being. Given these two conditions, we can state the following result: Let $m(x_i)$ be some increasing function of x_i. Then the only function $d(x_i, y_i)$ that satisfies the permutation and monotonicity conditions is one that can be put in the form $d(x_i, y_i) = m(x_i) - m(y_i)$.

Consider the case that was represented by the permuation condition:

(1) $d(x_i, y_i) + d(x_j, y_j) = d(x_j, y_i) + d(x_i, y_j)$.

Let us determine the nature of the function $d(x_i, y_i)$ when we hold the values of the variables x_j, y_j constant. In particular, let $x_j = y_j = k$, where k is some constant. Because of part (a) of the monotonicity condition, we know that for the case where $x_j = y_j$, $d(x_j, y_j) = 0$. We can, therefore, rewrite (1) as follows:

(2) $$d(x_i, y_i) = d(k, y_i) + d(x_i, k).$$

Let $g(x_i) = d(x_i, k)$ and $h(y_i) = d(k, y_i)$. Then we will have

(3) $$d(x_i, y_i) = g(x_i) + h(y_i).$$

No matter what the nature of $d(x_i, y_i)$, it will always be possible to define it in terms of the functions represented in (3). Now assume that $x_i = y_i$. As before, we must have $d(x_i, y_i) = 0$ in this case. This entails, given (3), that $g(x_i) = - h(y_i)$, or, in other words, since x_i and y_i are equal, that $h(y_i) = - g(y_i)$. Thus,

(4) $$d(x_i, y_i) = g(x_i) - g(y_i).$$

Finally, since $g(x_i)$ must be an increasing function of x_i on account of part (b) of the monotonicity condition, we can rewrite (4) as

(5) $$d(x_i, y_i) = m(x_i) - m(y_i).$$

This establishes our result.

Given (5), we see that we have once again returned to Fleming's result. For moral desirability, as defined by the MDF, must now be an additive function of measures of individual well-being. Since $d(x_i, y_i) = m(x_i) - m(y_i)$, we have that $d_1(x, y) + d_2(x, y) + \ldots + d_n(x, y) = m(x_i) - m(y_i) + m(x_2) - m(y_2) + \ldots + m(x_n) - m(x_n) - m(y_n)$, and therefore that

(a) $x\text{P}y$ iff $\Sigma_i m(x_i) > \Sigma_i m(y_i)$;

(b) $x\text{I}y$ iff $\Sigma_i m(x_i) = \Sigma_i m(y_i)$;

(c) $y\text{P}x$ iff $\Sigma_i m(x_i) < \Sigma_i m(y_i)$.

Thus, the aggregative existence function, plus some weak ethical constraints, will also generate Fleming's result. Of course, in one sense, Fleming's result seems stronger, since the ordinal representation assumption looks to be weaker than the aggregative existence function. But from another point of view, the two results are essentially equivalent, in that the ordinal representational assumption can only be justified in Fleming's context if something like the aggregative existence assumption is true. But it is clear, given the difference principle, that either one assumption or the other must receive further justification before we can conclude that utilitarianism has returned to firm conceptual ground. Let us, therefore, turn to the Harsanyi result to see if the addition of uncertain prospects to the situation under evaluation will further clarify the status of either the ordinal representation assumption or the aggregative existence assumption.

6

As we've seen, one of the features of Fleming's account was that it made no assumptions about the actual manner in which individual well-being or social welfare was to be measured. All Fleming wanted to show was that there must exist certain relations between social welfare and individual well-being if some weak ethical postulates were satisfied. Towards the end of his paper, though, Fleming suggested that the work of von Neumann–Morgenstern on the cardinalization of utility with respect to behavior under uncertainty might be relevant to not only the problem of measuring individual well-being, but also to that of measuring social welfare.[14] In his 1955 paper, 'Cardinal Welfare, Individualistic Ethics, and Interpersonal Comparisons of Utility', Harsanyi develops the implications of this suggestion and produces a powerful synthesis of the work of Fleming and von Neumann–Morgenstern. In our discussion of Harsanyi's result, we shall follow, with some modifications, Harsanyi's later statement of his theorem in his 1977 paper, 'Bayesian Decision Theory and Utilitarian Ethics'.[15]

Assume that in the case of risk and uncertainty, an individual's choices can be represented as choices among different lotteries whose prizes represent the individual's position in different states of the world, where "each position is characterized by finitely many economic and noneconomic variables. ." (p. 223). If there are r-different variables characterizing these positions, then any position can be regarded as a point in an r-dimensional Euclidean space E^r. Then, any lottery can be described as $L = (A_1/e_1; \ldots, A_k/e_k; \ldots, A_K/e_K)$, where the lottery L yields prizes $A_1, \ldots, A_K$ depending upon which one of K mutually exclusive and exhaustive events $e_1, \ldots, e_K$ occurs. This lottery will be said to be *risky* or *uncertain* depending upon whether the probability $p_k = \text{Prob}(e_k)$ associated with the event e_k is objectively defined by a process in the world or subjectively assigned by the agent. What von Neumann–Morgenstern showed was that if the individual's choices among a set of lotteries satisfied certain Bayesian rationality postulates, a cardinal utility function U_i will be defined for the individual which will assign to any lottery L a utility equal to its *expected utility*, where

$$U_i(L) = \Sigma_k p_k U_i(A_k).$$

The individual will thus act as if he were trying to maximize U_i. Any utility function U_i having this expected utility property will be called a *von Neumann–Morgenstern utility function*, (vN–M).

Since Harsanyi regarded the Bayesian rationality postulates as essential requirements for rational behavior, he found it natural to require that both individual and social preferences should satisfy them. If we add a strong Pareto requirement delimiting the nature of the relationship between individual and social preferences, we have Harsanyi's complete set of postulates:

POSTULATE A. The social welfare function is a vN–M utility function W.

POSTULATE B. Each individual's preferences are represented by a vN–M utility function U_i.

POSTULATE C. Suppose that there exists at least one individual i such that $x\mathrm{P}_i y$ and no individual j such that $y\mathrm{P}_j x$. Then the social preference will be $x\mathrm{P}y$.

With these postulates, Harsanyi was able to prove the following theorem:

THEOREM H. W is a weighted sum of the individual utilities, of the form $W = \Sigma a_i U_i$, with $a_i > 0$ for $i = i, 2, \ldots, n$.

This result is similar in structure to the Fleming result, for it establishes that social welfare must be an additive function of some cardinal measure of individual well-being. In order to transform the social welfare function into classical utilitarian form, Harsanyi notes that his postulates (as well as Fleming's) must be supplemented by one additional postulate:[16]

POSTULATE D. If all individuals' utility functions $U_1, U_2, \ldots, U_n$ are expressed in equal utility units, then the social welfare function W must assign the same weight to all these utility functions.

This postulate, which Harsanyi calls the requirement of the equal treatment of all individuals, will require that $a_1 = a_2 = \ldots = a_n = a$.

Since Harsanyi's postulate A requires that social preferences be defined by a cardinal social welfare function, we might be inclined to suspect, in the light of our discussion of Fleming, that it is this postulate which is critical for Harsanyi's result. After all, we've already seen the power lurking in the weaker assumption of the existence of an ordinal social welfare function. Of course, Harsanyi's assumption of the social welfare function has a much different basis than the comparable assumption in Fleming. Fleming's social welfare function followed from the ordinal representational assumption, while that of Harsanyi follows from the assumption that social preference should satisfy the Bayesian rationality postulates. Still, one might argue, as does Diamond in a review of Harsanyi's 1955 paper, that individual choice and social choice are two different things, with the implication that requirements that may be reasonable for individual choice will not be reasonable when transferred to the context of social choice.[17] Thus, Diamond argues, for example, that the so-called sure thing principle, while reasonable for individual behavior, may result in the 'unfair' treatment of individuals when applied to social choice.[18] While some objections of this type might be valid, they will in fact miss the real basis of Harsanyi's theorem. For as we will show, it is possible to restate Harsanyi's result in a very natural way which omits postulate A entirely. This restatement will show that the real source of Harsanyi's result is the requirement that social preference be defined on the basis of individual vN–M utility functions.

7

Prior to the work of von Neumann–Morgenstern, as we noted at the beginning of the paper, the concept of cardinally measurable individual utility had fallen into disfavor, since a strictly ordinal concept of utility seemed sufficient for describing rational behavior under certainty. Even when individual behavior could be described by a real-valued utility function, the numbers assigned by such a function were largely arbitrary, being unique only up to a *monotone transformation*. In other words, given a set of numbers representing an individual's utility among a group of prospects, any other set of numbers would do equally well, providing they defined the same *ordering* of these prospects. Thus, assuming an individual preferred prospect x to y, y to z and x to z, we could assign any triplet of numbers U_x, U_y, U_z to represent these utilities, providing $U_x > U_y > U_z$. Unless some procedure could be found for comparing *utility differences*, it was not possible to narrow the system of transformations up to which these numbers were defined.

While behavior under certainty furnished no way out of this impasse, von Neumann and Morgenstern showed how behavior under uncertainty could. For suppose we knew that an individual was indifferent between the sure prospect of y and the lottery offering a 50% chance of either x or z. If we define the individual's utility function as satisfying the expected utility property, then we will obtain the following result:

$$U_y = \tfrac{1}{2}U_x + \tfrac{1}{2}U_z.$$

Now if we multiply both sides of this equation by 2 and rearrange its terms, we will obtain:

$$U_x - U_y = U_y - U_z.$$

Given this relationship, it will no longer be the case that any triplet of numbers satisfying the relation $U_x > U_y > U_z$ will do. Rather, any two sets of numbers will do, provided that they are related by a *linear transformation*. Thus, if U_x is number representing the utility of x, then so is any number U_x' with the property $U_x' = a + bU_x$, where a, b are constants and $b > 0$. Utilities which are unique up to such positive linear transformations are *cardinal utilities*, and they are said to be unique up to the choice of a zero point (the constant a) and a unit of measurement (the constant b). By formulating a set of consistency requirements for behavior under uncertainty, the so-called Bayesian postulates, which entailed the existence of a utility function with the expected utility property, von Neumann and Morgenstern thus defined a measure of utility that was cardinal, i.e., unique up to linear transformations.[19]

Even though vN–M utilities are cardinal, of course, there is still quite a bit of looseness in the manner in which the utility numbers are correlated with prospects. For any given set of utility numbers, we can obtain an infinite number of equally good sets of numbers, provided that each set is related to every other by some linear transform. The arbitrariness of any particular assignment of cardinal utility numbers is no problem for the case of an individual's utility function, since any set out of the transformation group will describe his behavior under uncertainty equally well. However, there may very well be a problem if we wish to compare the utilities of different individuals. For if we have no basis for comparing the zero points or the units of the different utility functions, trying to compare the utility numbers for different individuals would be just as meaningless as trying to construct an unknown object from a set of blueprints of its components that had no points of reference or units of scale. Thus, suppose, for example, that we knew the following vN–M utilities for individuals i and j: x_i, y_i, x_j, y_j and that $x_i > y_i$ and $y_j > x_j$. What could we conclude from the fact that $x_i > y_j$ and $x_i - y_i > y_j - x_j$, for example? If we knew nothing about the zero points or units of the two different utility functions, the answer would be absolutely nothing. The fact that $x_i > y_j$ would convey no information about the relative well-beings of individual i in situation x or j in situation y. Nor would the fact that $x_i - y_i > y_j - x_j$ tell us anything about the relative 'intensities' of preferences. Since this kind of information is important for ethical judgements, the problem of *interpersonal utility comparisons* would seem to be a serious impediment to the ethical use of vN–M utilities.[20]

Since Harsanyi's postulates A to C made no mention of interpersonal comparisons, it might appear that his result somehow by-passed this problem. This is not the case, however, for the social welfare function which followed from these postulates had the form $W = \Sigma a_i U_i$, with $a_i > 0$ for $i = 1, 2, \ldots, n$. If no interpersonal comparisons of utility were possible, then the constants a_i would have to be assigned in a completely arbitrary manner. This would in turn have the consequence that the social preference for situation x over y, for example, would have no relation to the issue of whether situation x contained more individual well-being than y, or even more *von Neumann–Morgenstern utility* than y. This is obviously a far cry from the classical conception of utilitariansim. Unless some kind of interpersonal comparison of utility is postulated, it appears that Harsanyi's neo-utilitarian result is utilitarian in name only.

Harsanyi, of course, does remedy the above problem, for his addition of postulate D to the first three brings his result securely back into the utilitarian fold. For postulate D requires the social welfare function to be a symmetric function of the individual utility functions if these functions are *expressed in the same unit.* Without this assumption, any set of individual utilities U_i could be replaced by a new set U_i', providing, as we've seen, $U_i' = a_i + b_i U_i$, where a_i, b_i are constants and $b_i > 0$. And these constants could vary from individual to individual. But postulate D limits the set of possible transformations further, for it requires that any b_i be the same for all individuals. Thus, the allowable transformation becomes $U_i' = a_i + bU_i$, where a_i, b are constants and $b > 0$. The constant a_i may still vary among individuals, but b must be held constant across individuals. Given the potential variability of individual vN–M utility functions, postulate D plays an important role in limiting the large class of utilitarian-like welfare functions allowed by Harsanyi's theorem H to one's more properly called utilitarian. But as we shall now see, it is possible to reformulate Harsanyi's result in such a way that if we omit postulate A the combination of postulates B and the equal unit assumption of postulate D will be the decisive factor in the derivation of a utilitarian social welfare function.

8

We have seen that in the case of Fleming's result, the assumption that judgements of social preference could be represented in a certain form would, when combined with some weak ethical postulates, entail that measures of individual well-being must themselves have a certain form and relation to social welfare. The existence of a social welfare function that satisfied Fleming's postulates would require a cardinally measurable conception of individual well-being. In a sense, Harsanyi's approach may viewed as turning the Fleming result on its head. For we can show that if individual well-being is assumed to be measured by cardinal vN–M utility functions of a common unit, social preference must be determined by a function that defines social welfare in a cardinal manner.

Let us assume, as in Harsanyi's postulate B, that individual well-being is measured by vN–M utility funcitons U_i with x_i, y_i representing individual i's vN–M utility for situations x and y. Moreover, we shall make the assumption contained in postulate D that all individual utility functions are expressible in a common unit, so that any U_i will be unique up to the transform $U_i' = a_i + bU_i$, with a_i, b constants and $b > 0$. Thus, we may restate the new postulate B as follows:

POSTULATE B′. Each individual's preferences are represented by a vN–M utility function U_i that is defined in terms of a common unit.

Given the fact that any utility function U_i is unique only up to certain linearity preserving operations, it seems reasonable to require that the social preference should be independent of operations which change the numerical values of U^i but preserve their significance as utility functions defined up to certain transformations. If $R(x, y)$ represents the social preference between x and y and $R_i(x, y)$ the utility measures associated with the individual's well-being in situations x and y, then we can replace Harsanyi's postulate A with the following:

POSTULATE A′. If for all i, $R_i(x', y')$ is the result of a significance-preserving transformation of $R_i(x, y)$ and $R(x', y')$ the social preference for the situation which results from the set of all such transformations, then $R(x, y) \leftrightarrow R(x', y')$.

We may call this postulate the binary transformational independence condition.

Finally, to Harsanyi's postulate C requiring strong Pareto let us add the impartiality condition we used earlier in reformulating Fleming's result. This condition will require that the social preference will be independent of the particular individual with whom a given measure of well-being is associated:

POSTULATE C′. (a) Suppose there exists at least one individual i such that $x_i > y_i$ and no individual j such that $y_j > x_j$. Then $x\text{P}y$; (b) Suppose for all i, $x_i = y_i$. Then $x\text{I}y$; (c) Suppose x' is the situation which results from permuting the utility measures x_i among individuals. Then $R(x, y) \leftrightarrow R(x', y)$.[21]

Given this restatement of Harsanyi's postulates, we can how prove a comparable result:[22] For any situations x, y, construct the situations x^1, y^1 as follows: For x_1 and y_1, let $x_1^1 = x_1$ and $y_1^1 = y_1$. Then, for all individuals $i > 1$, let $x_i^1 = y_{i-1}^1$ and $y_i^1 = x_i^1 - x_i + y_i$, with the values of x_i^1 and y_i^1 being assigned first for $i = 2$, $i = 3$, and so on through $i = n$. Then, for all i, it will be the case that $(x_i - y_i) = (x_i^1 - y_i^1)$, which entails that $R_i(x^1, y^1)$ was obtained from $R_i(x, y)$ by the operation of adding or subtracting a constant a_i. Since this is an operation which preserves the significance of the utility measures x_i, y_i, postulate A′ will require that $R(x, y) \leftrightarrow R(x^1, y^1)$. Also, notice that $\Sigma_i(x_i - y_i) = \Sigma_i(x_i^1 - y_i^1)$ follows from the manner in which x^1 and y^1 have been defined. Now construct situation x^2 from x^1 as follows: For $i = 1$ through $i = n - 1$, let $x_i^2 = x_{i+1}^1$, and for $i = n$, let $x_n^2 = x_1^1$. Since x^2 was obtained from x^1 by permuting the utility measures x_i^1 among individuals, postulate C′, part (c) will require that $R(x^1, y^1) \leftrightarrow R(x^2, y^1)$. But notice that for all $i \neq n$, $x_i^2 = y_i^1$, with the result that $x^2\text{P}y^1$ iff $x_n^2 > y_n^1$, $x^2\text{I}y^1$ iff $x_n^2 = y_n^1$, and $y^1\text{P}x^2$ iff $x_n^2 < y_n^1$, because of parts (a) and (b) of C′. Finally, notice that the following relations hold:

(i) $\Sigma_i(x_i^2 - y_i^1) > 0$ iff $x_n^2 > y_n^1$ iff $x^2\text{P}y^1$;

(ii) $\Sigma_i(x_i^2 - y_i^1) = 0$ iff $x_n^2 = y_n^1$ iff $x^2\text{I}y^1$;

(iii) $\Sigma_i(x_i^2 - y_i^1) < 0$ iff $x_n^2 < y_n^1$ iff $y^1\text{P}x^2$.

Thus, since $\Sigma_i(x_i^2 - y_i^1) = \Sigma_i(x_i^1 - y_i^1)$ by inspection, and $\Sigma_i(x_i^1 - y_i^1) = \Sigma_i(x_i - y_i)$ from above, and $R(x^2, y^1) \leftrightarrow R(x^1, y^1) \leftrightarrow R(x, y)$, the result will be that

(a) $x\text{P}y$ iff $\Sigma_i(x_i - y_i) > 0$;

(b) $x\text{I}y$ iff $\Sigma_i(x_i - y_i) = 0$;

(c) $y\text{P}x$ iff $\Sigma_i(x_i - y_i) < 0$.

Let any social welfare function that satisfies properties (a) – (c) be *utilitarian.* Then we have established the following theorem:

THEOREM H′. A social welfare function satisfying postulates A′ – C′ is utilitarian.

Since this result is equivalent to that of Harsanyi when postulate D is added, we have essentially reproduced his result without using his postulate A.

9

In the light of theorem H′, it should be clear that just as the manner in which the concept of social welfare is defined will limit the nature of morally acceptable measures of individual well-being, so too will the manner in which individual well-being is defined limit the nature of a morally acceptable concept of social preference. If we assume that vN–M utility functions provide an appropriate way of measuring the well-being of individuals and that a common unit of measurement may be established for comparing the utility functions of different individuals, these assumptions will limit the class of minimumly acceptable ethical systems to just one, classical utilitarianism. Since the requirement that there exist a common unit of measurement is one that we would want to impose on any method of measuring individual well-beings that claimed for itself cardinal significance and ethical relevancy, the decisive factor in this derivation of utilitarianism must be the assumption of the moral appropriateness of vN–M utility functions. Since in Harsanyi's framework these utility functions are themselves the result of the assumption of the Bayesian rationality postulates, we must now turn from the issue of the nature of social decision-making to the question of rationality on an individual level.

It is Harsanyi's contention that the behavior of any individual who is 'rational' will satisfy the Bayesian postulates, which are "absolutely inescapable criteria of rationality." On the assumption that the individual's behavior is consistent enough to be representable by a well-defined ordinal decision function, Harsanyi lists two rationality postulates for behavior under risk and uncertainty:[23]

PROBABILISTIC EQUIVALENCE. Let

$$L = (A_1/e_1; \ldots; A_K/e_K) \text{ and } L^* = (A_1/e_1^*; \ldots; A_K/e_K^*)$$

and suppose that the decision maker *knows* the objective probabilities associated with events $e_1, \ldots, e_K$ as well as with events $e_1^*, \ldots, e_K^*$, and knows that these probabilities satisfy

$$\text{Prob}(e_k) = \text{Prob}(e_k^*) \text{ for } k = 1, \ldots, K.$$

Then he will be *indifferent* between lotteries L and L^*.

This postulate requires that the individual will be indifferent between two risky lotteries if they yield the same prizes with the same probabilities, even though the lotteries use different processes to generate these probabilities.

SURE-THING PRINCIPLE. Suppose that $A_k^* >_i A_k$ for $k = 1, \ldots, K$. Then $(A_1^*/e_1; \ldots; A_K^*/e_K) \gtrsim_i (A_1/e_1; \ldots; A_K/e_K)$.

Other things being equal, this postulate requires that the individual *not* prefer a lottery yielding less desirable prizes to one yielding more desirable prizes. Harsanyi points out that this requirement is the analog of the requirement in game theory that an individual avoid using any dominated strategies. It also has strong ties to the Pareto requirements that we have employed earlier in the paper.

While there are several other systems of postulates that will entail the existence of vN–M utilities, the postulates for risk and uncertainty employed by Harsanyi have the advantage of being simple and intuitive.[24] But no matter what the axiom system, all make use of some form of the sure-thing principle, which has also been called the strong independence condition. As a result, when the axiom systems are criticized, it is almost always the sure-thing principle which is the center of controversy, since it is this principle which is viewed as decisive for the cardinality of utility, as opposed to its mere ordinality.[25] But with the notable exception of Georgescu-Roegen and Chipman, the assumption of the existence of an ordinal utility function has gone unchallenged.[26] As we shall see, it is this assumption upon which the whole edifice of cardinal utility ultimately rests, much as the possibility of utilitarianism in Fleming's case ultimately came down to the existence of an ordinal social welfare function. Thus, as in Fleming's case, the prospects for cardinality will be dependent on the appropriateness of the ordinal representation assumption in the context of decision-making under consideration.

Within Harsanyi's framework, the existence of an ordinal utility function follows from the assumption of two postulates, which Harsanyi calls the *basic utility axioms*. Suppose $S_i(A)$ is the set of all positions that individual i prefers to A or regards as equivalent to A, while $I_i(A)$ is the set of all positions that i finds to be less preferable than A or equivalent to A. Then the rational behavior of individual i under *certainty* must fulfill the following postulates:

COMPLETE ORDERING. The nonstrict preferences of individual i define a complete ordering of all possible positions.

CONTINUITY. For individual i, both his superior set $S_i(A)$ and his inferior set $I_i(A)$ for any particular position A are closed subsets of the r dimensional Euclidean space E^r in which these positions can be represented.

If these postulates are satisfied, then, as Debreu has shown, the individual's behavior will be equivalent to maximizing an ordinal utility function.[27]

Since an ordinal utility function assigns a real number to each possible position, with the property that one position will be preferred to another if its utility number is higher, the operative assumption once again is the ordinal representation assumption. Unlike the case of Fleming, however, the existence of a numerical representation of an ordering follows logically from an explicitly stated postulate, namely, the continuity postulate. If the continuity postulate is indeed a requirement of rationality, then the ordinal representation assumption will be true, at least in the context of individual decision-making.

In order to see the implications of the continuity assumption, let us consider how it would be applied in the situation where there are only two commodities relevant to the individual's preference. Let the x-coordinate represent the amount of commodity x an individual could receive and the y-coordinate the amount of commodity y. Suppose that there exists a point on the y-axis, A', and a point on the x-axis, A'', with the property that the curve $A'A''$ connecting these points and the point A is the locus of all points (x, y) equivalent to point A for the individual. Then, we may call the curve $A'A''$ an *indifference curve*, and the inferior set $I(A)$ will be the set consisting of all points below $A'A''$ plus all points lying on $A'A''$. Similarly, the superior set $S(A)$ will consist of all points above $A'A''$ plus all points on $A'A''$. The individual's behavior as represented by this situation will satisfy the continuity assumption, since both the sets $I(A)$ and $S(A)$ will be *closed*, that is, will contain *all of their boundary points*, namely, the curve $A'A''$.[28]

But now suppose that the curve $A'A''$ has a different interpretation than it had above. As before, any point on the curve will be preferred to any point below the curve, but now assume that one point on the curve will be preferred to another point on the curve if it has a higher x-value than the latter. The curve $A'A''$ will no longer be an indifference curve, but rather will be what Little, for example, calls a *behavior curve.*[29] And it will no longer be the case that the sets $I(A)$ and $S(A)$ are closed, since these sets will no longer contain their boundary points. Thus, for example, the set $I(A)$ will now include all points below $A'A''$ and all points on $A'A''$ *above* point A. But since it will not include those points on $A'A''$ lying below A, the region *below* this segment of $A'A''$, though in $I(A)$, will be *unbounded*. Since behavior described by this kind of model will not satisfy the continuity assumption, it will not qualify as 'rational' behavior, even though it will define a complete ordering of all points in this space, if we assume a set of comparable behavior curves. But it is hard to see what is irrational about this kind of behavior beyond the fact that it is not representable by a real-valued funcion.

Since the above example might be too abstract for identifying the nature of the difficulty with the continuity assumption, let us consider what kind of motivational factors could produce the behavior curve described above. Suppose the individual was concerned with satisfying two distinct sets of needs, N^1 and N^2, and that one set, N^1, had priority over the other, N^2. Thus, the individual would always try and maximize the satisfaction of N^1 before he would concern himself with N^2. Now suppose the curve $A'A''$ represented those combinations of commodities x and y that produced equal levels of satisfaction of N^1 and that commodity x was directly correlated with the satisfaction of N^2, so that the more x the individual possessed, the higher the satisfaction of N^2. In this case, $A'A''$ would represent precisely the behavior curve described above. For any point A, the inferior set $I(A)$ would consist of all points below $A'A''$ and all points on $A'A''$ above point A, and the superior set $S(A)$ would consist of all points above $A'A''$ and all points on $A'A''$ below point A. And if we assume that there exists a set of behavior curves with the property that any point on a curve represents equal satisfaction of N^1, then the individual's behavior will define a complete ordering of all possible combinations of commodities x and y. Clearly, this behavior is a paradigm example of rational behavior, even though it violates the continuity postulate.

Since the continuity assumption represents the ordinal representation assumption in the context of individual decision-making, the parallels between the present discussion and the earlier discussion in the Fleming case should be clear. Just as the ordinal representation assumption had the consequence of ruling out ethical systems that ordered social situations on the basis of the relative priority of individual claims, so too does it rule out individual behavior that orders possible positions on the basis of the relative priority of individual needs. In effect, what the ordinal representation assumption requires is that all dimensions relevant to a choice should be collapsible into one dimension. It thus ignores the possibility that there may very well exist from a rational point of view factors of choice that are *incommensurable*. In the case of social decision-making, the different types of claims that individuals make may not be morally conflatable into a common index of desirability, as the aggregative existence assumption requires. And in the case of individual decision-making, the different types of needs that constitute an individual's motivational system may not be rationally conflatable into a common index of utility, as the continuity assumption requires. It is surprising that the mere assumption of ordinal representability has this conflating property, but apparently the real number system is just not rich enough to capture the multidimensional nature of rational decision-making.

10

As Georgescu-Roegen points out in his important 1954 paper, 'Choices, Expectations, and Measurability', the notion that human needs and wants are hierarchized is not a new one.[30] After all, it is this hierarchy that is the essence of any argument explaining the well-known principle of marginal utility:

all these successive units have for their possessor an intensity of utility decreasing from the first unit which corresponds to the most urgent need to the last, after which satiety sets in (p. 194).

Georgescu-Roegen lists several principles that have appeared in classical economic literature in association with the notion of a hierarchy of wants. The first he calls the *principle of the subordination of wants*, which is the claim that the satisfaction of a lower order want permits the next higher order want to manifest itself. This principle in turn requires that all wants are satiable at some point, which is the *principle of satiable wants*. It is this latter principle, Georgescu-Roegen claims, that was rechristened by the founders of marginal utility theory as the *principle of diminishing marginal utility*. Moreover, since there always seems to be a *next* want after one is satisfied, we need a *principle of the growth of wants*, which assumes that absolute satiety never exists. Finally, there is the principle that seems to have escaped the attention of the neoclassical economists, namely, the *principle of the irreducibility of wants*. For some reason the neoclassicists moved from a theory of many wants to that of a single want to which all others could be reduced, namely, *utility*. But if there is no common denominator to all wants, then the attempt to define a real-valued utility function, whether ordinal or cardinal, will fundamentally misrepresent the nature of human motivation. Despite the fact that the want patterns of individuals will clearly differ to some degree, Georgescu-Roegen argues that these patterns will have much in common:

(1) *The hierarchy of wants seem to be for all men identical up to a certain rank.* One may be almost sure that this refers at least to: thirst, hunger, leisure, shelter. (2) *individuals belonging to the same culture are likely to have in common still a greater number of wants at the top of the hierarchy than those common to all men* (pp. 197–198).

While these are strong claims for an economist to be making, they receive independent support in the work of the pyschologist Maslow whose book *Motivation and Personality* was published in the same year as Georgescu-Roegen's article. For Maslow presents a theory of human motivation in his book which parallels the principles Georgescu-Roegen had enunciated. Claiming that his theory derives most directly from clinical experience, Maslow argues that human needs arrange themselves into a hierarchy of needs which he calls the *basic needs*.

The chief dynamic principle animating this organization is the emergence in the healthy person of less potent needs upon gratification of the more potent ones. The physiological needs, when unsatisfied, dominate the organism, pressing all capacities into their service and organizing these capacities so that they may be most efficient in this service. Relative gratification submerges them and allows the next higher set of needs in the hierarchy to emerge, dominate, and organize the personality, so that instead of being, e.g., hunger obsessed, it now becomes safety obsessed. The principle is the same for the other sets of needs in the hierarchy, i.e., love, esteem, and self-actualization (p. 59).

Maslow, like Georgescu-Roegen, points out that actual behavior is likely to be overdetermined, since there is no one-to-one correspondence between needs and goods. And Maslow adds that we must not assume that all behavior is determined by the basic needs, since not all behavior is determined.

Obviously, both work of Georgescu-Roegen and Maslow are, to some extent, highly conjectural; and both are probably a mixture of empirical *and* normative claims about the structure of human motivation. But just as we can investigate the implications of the normative model of rationality defined by the Bayesian rationality postulates, so too can we investigate the model of human motivation represented by the *basic need theory*. Certainly the basic need theory is a plausible enough theory of rationality to call into question the viability of the continuity postulate, whether or not the *actual* specification and ranking of needs suggested by Georgescu-Roegen or Maslow are correct. Let us assume, therefore, that some simplified version of the basic need theory is correct, so that, as Georgescu-Roegen claims, "choice aims at satisfying the greatest number of wants starting with the most important and going down their hierarchy" (p. 198). What will be the implications of this assumption for the neo-utilitarian theories we have been analysing?

If we assume the individual's rational preferences are directed at satisfying a hierarchy of needs $N^1, N^2, \ldots, N^L$, where N^1 is the most urgent need, N^2 the next urgent, and so on, then the position A' will be preferred to A'' if the highest-order need not satisfied equally well by A' and A'' is better satisfied by A' than by A''. These preferences will define a complete ordering of all possible positions in E^r, but will not, of course, satisfy the continuity postulate, as we have seen for the two need case. While indvidual i's behavior will not be representable by an ordinal utility function U_i, it will be representable in terms of lexically ordered utility vectors $(U_i^1, U_i^2, \ldots, U_i^L)$. Thus the individual may be viewed as first maximizing the ordinal utility U^1, then U^2, and so on, where each utility vector U^r is correlated with the satisfaction of need N^r. Then if we assume that individual i's behavior under risk and uncertainty will satisfy Harsanyi's postulates of probabilistic equivalence and the sure-thing principle, then, as Hausner has shown, the individual may be viewed as maximizing the expected value of $(\bar{U}_i^1, \bar{U}_i^2, \ldots, \bar{U}_i^L)$, where $\bar{U}_i^r$ is a utility vector corresponding to the need N^r which will be cardinally significant and satisfy the expected value property of vN–M utilities.[31]

If we replace the vN–M utility functions in our postulate B' with the corresponding utility vectors $(\bar{U}_i^1, \bar{U}_i^2, \ldots, \bar{U}_i^L)$ and make the assumption that for each $\bar{U}_i^r$ there exists a common unit of measurement for all individuals i, then a modified version of our theorem H' will follow. Let x_i^r represent the vN–M utility that individual i receives in position x with respect to the satisfaction of need N^r. Then xPy will be the social preference iff level k is the smallest level r such that it is not the case that $\Sigma_i(x_i^r - y_i^r) = 0$ and it is the case that $\Sigma_i(x_i^k - y_i^k) > 0$. While these social preferences will not be representable by a social welfare function, they will be representable by what we may call a *lexical social welfare function*, consisting of the lexically ordered vectors $(W^1, W^2, \ldots, W^L)$. We shall call this result *lexical utilitarianism.*[32]

While the lexical utilitarian principle does have some utilitarian-like structural properties to it, there is an important sense in which it is not utilitarian at all, but rather what we may call *Rawlsian*. For given the manner in which Rawls defines his lexical difference principle in *A Theory of Justice*, the assumption of a theory of basic needs has the consequence of collapsing utilitarianism and the lexical difference principle into a common theory of social evaluation. For recall that Rawls' difference principle does not tell us to maximize the prospects of the worst-off individual, but rather the worst-off *class*. Thus, to define this class, Rawls tells us the "we are to aggregate to some degree over the expectations of the worst-off . . ." (p. 98). But if we assume that the theory of basic needs is relevant to Rawls' theory, then the worst-off class will always be those individuals who have not satisfied their most basic needs, N^1, the next worst-off class those who have not satisfied N^2, and so on. The concept of basic needs thus provides a natural way of defining classes from the standpoint of the difference principle.[33] Indeed, even if it is not the case that human behavior satisfies the basic need theory in actual practice, there may be good arguments, from the standpoint of justice, for treating human behavior as if it did. Then, if we require the difference principle to maximize the satisfaction of N^1 needs across all the individuals in society, it will in effect maximize the aggregate well-being of the worst-off class. And the well-being of this class will have to be maximized before the aggregate well-being of the next worst-off class is considered, and so on. The result will be, then, that the lexical difference principle, as defined for the concept of basic needs, will be equivalent to the utilitarian principle defined for basic needs, that is, to lexical utilitarianism. Such a principle may properly be called Rawlsian, since it represents the basic Rawlsian injunction that the prospects of one class are never to be sacrificed to achieve greater prospects for a better-off class.

Given the language we introduced earlier in the paper, we may say that the lexical utilitarian principle combines the *claim-based* view of morality with the *aggregative* view. But the aggregative operation is not used to define an index of 'desirability' in terms of which we can rank different situations. Rather, it is employed instead to define the nature of an aggregate moral entity, the Rawlsian class, which is viewed as the representative of a particular moral claim. The social preference, however, will be determined on the basis of the highest ranked claim. In case of ties, the next highest ranked claim will be considered, and so on. The utilitarian aspect of the lexical utilitarian principle, therefore, has a secondary role from the point of view of the social preference. For the lexical utilitarian principle is fundamentally the principle of a claims-based ethical system.

On the basis of this paper's analysis, we have seen that there is an intimate relationship between the nature of individual and social decision-making. As the neo-utilitarians correctly surmised, the logical structure of an ethical system is ultimately determined by the logical structure of the conception of individual well-being upon which it is based. That this kind of connection exists should not be at all surprising, since ethical decision-making is in some sense merely a generalized form of individual decision-making. To a great extent, what is rational for an individual should also be rational from the standpoint of an ethics. Thus, if there exists a hierarchy of needs in terms of which human decision is or should be structured, then an ethical system must reflect this ordering in the form of a hierarchy of ordered claims. The ethical system which results is at best a distant relative to classical utilitarianism, which, as we have seen, presupposes a restrictive view of human motivation. Rather, it is a hybrid form of Rawls' difference principle, so modified as to take into account the hierarchical nature of needs. The neo-utilitarian approach has not demonstrated the bankrupcy of the Rawlsian approach nor the superiority of the classical utilitarian approach; rather, it has paved the way towards a synthesis of the best features of both.[34] The resulting theory may perhaps allow the two camps to better understand the true sources of their disagreement, which, if this paper is right, must ultimately be found in the theory of human rationality that each embraces.

Department of Philosophy, Stanford University

NOTES

[1] The new welfare economics began, of course, with Pareto and reached 'maturity' in works by Bergson (1938), Samuelson (1948, 1950) and Graaff (1957).

[2] Interestingly enough, von Neumann and Morgenstern made little actual use of the concept of utility in their work. In someways, their result was anticipated by the work of Ramsey (1931) on subjective degrees of belief.

[3] Harsanyi's classic paper is the 1955 'Cardinal Welfare, Individualistic Ethics, and Interpersonal Comparisons of Utility'. Later expositions of this result are found in Harsanyi (1975b, 1977a, 1977b, 1977c).

[4] Fleming's result actually came before the Harsanyi result, but made no use of the von Neumann–Morgenstern concept of utility. See Fleming (1952).

[5] The difference principle is discussed in Rawls (1971), pp. 75–83. For a scathing utilitarian counter-attack on the difference principle, see Harsanyi (1975a).

[6] In a sense, the classic statement of the new welfare economics is in Bergson (1938).

[7] For a discussion of the concept of ordering and its various axiomatic formulations, see Sen (1970), ch. 1*.

[8] The implications of this kind of interpersonal comparison were first studied by Suppes (1966).

[9] In fact, if $\widetilde{R}$ are the only kind of judgements of well-being that we allow, a form of Rawls' difference principle will be the only principle that can satisfy a modified set of ethical postulates. See Arrow (1977), d'Aspremont and Gevers (1977), and Strasnick (1975, 1979a).

[10] This particular formulation of the leximin principle is due to Sen (1979). For axiomatic analyses of the leximin principle, see Arrow (1977), d'Aspremont and Gevers (1977), Deschamps and Gevers (1978), Hammond (1976), Sen (1970, 1976, 1979), and Strasnick (1975, 1979a). For a discussion and derivation of the non-lexical form of the difference principle, see Strasnick (1976a).

[11] See Rawls (1971), p. 30 for a discussion of the contrast between deontological and teleological theories.

[12] See also Birkhoff (1948), especially his Theorem 2 or Ex. 6*, p. 32.
[13] For an analysis of the properties of an ethical system that is based on ordering the 'priority' of individual claims, see Strasnick (1975, 1976b, 1977).
[14] See Fleming (1952), pp. 260–267.
[15] The results presented by Harsanyi in his series of papers are for the most part identical, though some of the axioms are modified.
[16] See Harsanyi (1977c), p. 227.
[17] See Diamond (1976). Also, for a more detailed critique, see McClennan (1979).
[18] For Harsanyi's persuasive reply to Diamond's argument, see Harsanyi (1975b).
[19] For a good discussion of the von Neumann and Morgenstern approach, see Luce and Raiffa (1957). See Ellsberg (1954) for a contrast between the vN–M concept of utility and other 'classic' economic formulations of utility.
[20] See Sen (1970), chapters 7, 7* for a good discussion of this problem.
[21] Actually, a weaker impartiality condition and a weak Pareto condition could be used to derive postulate C' as a lemma, but this postulate seems weak enough for our purposes.
[22] This result is modified from Strasnick (1975), chapter 6.
[23] See Harsanyi (1977c). In Harsanyi (1955), the Marschak (1950) axioms are used instead.
[24] For other axiomatic accounts, see Friedman and Savage (1948), Herstein and Milnor (1953), Marschak (1950), and, of course, von Neumann and Morgenstern (1944).
[25] For some discussions of this axiom, see Friedman and Savage (1948), Luce and Raiffa (1957), McClennen (1979), and Samuelson (1952).
[26] See Georgescu-Roegen (1954) and Chipman (1960).
[27] See Debreu (1954).
[28] This example is from Harsanyi (1977a), p. 30.
[29] See Little (1950).
[30] Georgescu-Roegen quotes a nice example from Plato: "Now the first and chief of our needs is the provision of food for existence and life. The second is housing and the third is raiment and that sort of thing." *Republic* 369D.
[31] See Hausner (1954) and Thrall (1954).
[32] In some ways, the classical utilitarian Mill anticipated lexical utilitarianism in his discussion of justice in Mill (1962). Consider, for example, the following: "Our notion, therefore, of the claim we have on our fellow creatures to join in making safe for us the very groundwork of our existence, gathers feelings around it so much more intense than those concerned in any of the more common cases of utility, that the difference in degree becomes a real difference in kind. The claim assumes that character of absoluteness, that apparent infinity, and incommensurability with all other considerations, which constitute the distinction between the feeling of right and wrong and that of ordinary expediency and inexpediency." p. 310. Also, see p. 321.
[33] In Rawls' theory, well-being is not represented in terms of a subjective index of satisfaction, like utility, for example. Instead, Rawls uses an objective index measuring social primary goods, which "are things a rational man wants whatever else he wants." See Rawls (1971), pp. 90–95. But Rawls must still aggregate the different kinds of social primary goods ("rights and liberties, powers and opportunities, income and wealth") into a common index. If a theory of basic needs is true, a corresponding lexical ordering of social primary goods might be an appropriate analog for primary goods in Rawls' theory.

[34] For a discussion of the sense in which ethical theories may be both true and false, see Strasnick (1979b).

REFERENCES

Arrow, K., 'Extended sympathy and the possibility of social choice', *American Economic Review* 67 (1977), 219–225.

Bergson, A., 'A reformulation of certain aspects of welfare economics', *Quarterly Journal of Economics* 52 (1938), 310–334.

Birkhoff, G., *Lattice Theory*, American Mathematical Society, (1948).

Chipman, J. S., 'The foundations of utility', *Econometrica* 28 (1960), 193–224.

d'Aspremont, C. and Gevers, L., 'Equity and the informational basis of collective choice', *Review of Economic Studies* 46 (1977), 199–210.

Debreu, G., 'Representation of a preference ordering by a numerical function', in Thrall *et. al.* (1954), 159–165.

Deschamps, R. and Gevers, L., 'Leximin and utilitarian rules: a joint characterization', *Journal of Economic Theory* 17 (1978), 143–163.

Diamond, P., 'Cardinal welfare, individualistic ethics, and interpersonal comparisons of utility: a comment', *Journal of Political Economy* 75 (1967), 765–766.

Ellsberg, D., 'Classic and current notions of measurable utility', *Economic Journal* 64 (1954), 528–556.

Fleming, M., 'A cardinal concept of welfare', *Quarterly Journal of Economics* 66 (1952), 366–384. Reprinted in Phelps (1973). Page references are to this edition.

Friedman, M. and Savage, L. J., 'The utility analyses of choices involving risk', *Journal of Political Economy* 56 (1948), 279–304.

Georgescu-Roegen, N., 'Choices, expectations, and measurability', *Quarterly Journal of Economics* 73 (1954), 503–534. Reprinted in Georgescu-Roegen, N., *Analytical Economics*, Harvard University Press, 1967. Page references are to this edition.

Graaff, J. Dev., *Theoretical Welfare Economics*, Cambridge University Press, 1957.

Hammond, P., 'Equity, Arrow's conditions, and Rawls' difference principle', *Econometrica* 44 (1976), 793–804.

Harsanyi, J., 'Cardinal welfare, individualistic ethics, and interpersonal comparisons of welfare', *Journal of Political Economy* 63 (1955), 309–321. Reprinted in Phelps (1973). Page references are to this edition.

Harsanyi, J., 'Can the Maximin Principle Serve as a Basis for Morality?' *American Political Science Review* 69 (1975a), 594–606.

Harsanyi, J., 'Nonlinear social welfare functions: do welfare economists have a special exemption from Bayesian rationality?' *Theory and Decision* 6 (1975b), 311–332.

Harsanyi, J., *Rational Behavior and Bargaining Equilibrium in Games and Social Situations*, Cambridge University Press, 1977a.

Harsanyi, J., 'Morality and the theory of rational behavior', *Social Research* 44 (1977b), 623–656.

Harsanyi, J., 'Bayesian decision theory and utilitarian ethics', *Working Paper* CP-404, Center for Research in Management Science, University of California, Berkeley, 1977c. Published in *American Economic Review* 68, 223–228. Page references are to this edition.

Hausner, M., 'Multidimensional utilities', In Thrall *et. al.* (1954), 167–180.
Herstein, I. N. and Milnor, J. W., 'An axiomatic approach to measurable utility', *Econometrica* 21 (1953), 291–297.
Little, I. M. D., *A Critique of Welfare Economics*, Oxford University Press. 1950.
Luce, R. D. and Raiffa H., *Games and Decisions,* John Wiley & Sons, New York, 1957.
Marschak, J., 'Rational Behavior, Uncertain Prospects, and Measurable Utility', *Econometrica* 18 (1950), 111–141.
Maslow, A., *Motivation and Personality,* Harper & Row, New York, 1954.
McClennan, E., 'Constitutional choice: Rawls vs. Harsanyi', 1979. Forthcoming in *Philosophy and Public Affairs.*
Mill, J. S., 'Utilitarianism', in M. Warnock (ed.), *John Stuart Mill*, Meridian Books, 1962.
Neumann, von J. and Morgenstern O., *Theory of Games and Economic Behavior*, Princeton University Press, 1944.
Phelps, E. S., *Economic Justice*, Penguin Books, 1973.
Ramsey, F., 'Truth and probability', in *The Foundations of Mathematics and Other Logical Essays*, Harcourt, Brace, & Co, 1931.
Rawls, J., *A Theory of Justice*, Harvard University Press, 1971.
Samuelson, P. A., *Foundations of Welfare Economics*, Harvard University Press, 1948.
Samuelson, P. A., 'The evaluation of real national income', *Oxford Economic Papers*, January, 1950. Reprinted in Phelps (1973).
Samuelson, P. A., 'Probability, utility, and the independence axiom', *Econometrica* 20 (1952), 670–678.
Sen, A. K., *Collective Choice and Social Welfare*, Holden Day, 1970.
Sen, A. K., 'Welfare inequalities and Rawlsian axiomatics', *Theory and Decision* 7 (1976), 243–262.
Sen, A. K., 'Social choice theory', to appear in K. J. Arrow and M. Intriligator, (ed.), *The Handbook of Mathematical Economics,* (1979).
Strasnick, S., *Preference Priority and the Maximization of Social Welfare*. PhD Thesis, Harvard University, 1975.
Strasnick, S., 'Social choice and the derivation of Rawls' difference principle', *Journal of Philosophy* 73 (1976a), 85–99.
Strasnick, S., 'The problem of social choice: Arrow to Rawls', *Philosophy and Public Affairs* 5 (1976b), 241–273.
Strasnick, S., 'Ordinality and the spirit of the justified dictator', *Social Research* 44 (1977), 668–690.
Strasnick, S., 'Extended Sympathy Comparisons and the Basis of Social Choice', *Theory and Decision* 10 (1979a), 311–328.
Strasnick, S., 'Moral structures and axiomatic theory', *Theory and Decision* 11 (1979b), 195–206.
Suppes, P., 'Some formal models of grading principles', *Synthese* 6 (1966), 284–306.
Thrall, R. M., 'Applications of multidimensional utility theory', in Thrall *et. al.* (1954), 181–186.
Thrall, R. M. *et. al., Decision Processes*, John Wiley & Sons, New York 1954b.

E. F. McCLENNEN

CONSTITUTIONAL CHOICE: RAWLS VS. HARSANYI*

1. INTRODUCTION

In recent years there has been considerable exploration of two approaches to the problem of the justification of principles for the ordering of social alternatives. One, associated with Rawls, but anticipation of which is to be found in Harsanyi, turns on the notion of what the representative person would choose given a state of (radical) uncertainty.[1] The other, to which Harsanyi has also contributed, but which is familiar to most through the work of Arrow, concerns the delimitation of a social welfare function which takes a set of individual rankings (or utilities) into a social ranking (or index of social utility). The notion, in this case, is to explore reasonable constraints which can be placed upon such a function, and to determine whether, given a suitable set of such constraints, the function can be uniquely defined.[2]

The history of the second research program is dominated by an important negative result, the well-known Arrow impossibility theorem for social welfare functions. In recent years, however, there have been a number of significant developments, which appear to promise a way out of the Arrow impasse. Essentially what is involved in these new results is that it is no longer assumed, as in Arrow's original formulation of the problem, that what is to be taken into a social ordering is a set of non-comparable individual orderings (what Arrow termed a 'profile'). By 'enriching' the information available concerning individual orderings in such a manner as to permit some form or other of interpersonal comparison (either in the form of ordinal comparability of levels of preferences or comparability of preference differences) the original Arrow impasse is avoided, and constructive results are obtained.[3]

Somewhat less familiar is the fact that the first research program has also been dominated by a parallel negative result. In this case, however, proposals for avoiding the problem have been around since the early Fifties. In essence, each such proposal amounts to a recommendation that one drop one or more of the conditions which, collectively, are inconsistent with the assumption of the existence of a decision rule for choice under uncertainty. Perhaps the most interesting (although by no means the only) results to emerge from this sort of relaxation of one or more of the conditions are (1) a rule

J. C. Pitt (ed.), Philosophy in Economics, 93–109.

to the effect that one should evaluate alternative uncertain prospects in terms of their (unweighted) average return (which is formally equivalent to the Laplacean principle of treating all uncertain events as equiprobable, and computing expected return against this equiprobable distribution), and (2) a variety of rules which stress exclusively considerations of minima and/or maxima.[4] Applying such an uncertainty model of decision making to the choice of a rule for ordering social states, we have Harsanyi, on the one hand, arguing vigorously for an average utility approach, and Rawls, on the other hand, defending a rule based exclusively on consideration of minima – the lexicographical minimax rule.

It will perhaps come as a surprise to those who are not familiar with the recent work on the definition of a social welfare function, that the two principal lines of positive results are: (1) a rule to the effect that the optimal social choice is that alternative (or set of alternatives) which maximizes the (unweighted) average utility of the participants; and (2) a rule which defines the optimal set in terms of what maximizes the minimum (and, in case of ties, breaks them by reference to maximizing the next level up, etc.) – in short, a lexicographical version of the minimax rule.

We have, then, strikingly similar results concerning appropriate rules for ordering social choices, whether we approach the problem of justification from the perspective of what the representative person would find it rational to propose under conditions of (radical) uncertainty, or from the perspective of what emerges from the imposition of 'reasonable' conditions on the definition of a social welfare function which aggregates individual interests or utilities. But we are also faced with the paradox that, in each case, the results are not univocal: that radically different approaches can be defended by reference to such constructions or models.

Put in a more historical perspective, the new results which have emerged throw us right back into the middle of the debate of the last century over utilitarianism vs. non-utilitarian theories. There is, moreover, a sort of irony about all of this. It was Rawls' attempt, some years ago, to develop a full-blown systematic alternative to a utilitarian theory of justice which perhaps did more than anything else to stimulate this whole double area of research (into uncertainty choice, and social choice). Yet the upshot has been that that appeal to conditions of rational choice has revitalized the whole utilitarian tradition: and those who find it attractive have been able, point by point, to counter Rawls' attempt to construct an alternative. With an eye to views which have yet to be published, one can speak here, in fact, of Rawls' attempts, for he has recently argued that the social welfare function approach is perhaps more convincing than the approach which he took in his book, *A Theory of Justice*, which turns on the notion of choice under conditions of (radical) uncertainty.[5] However, this strategic shift on Rawls' part is hardly to any avail: Harsanyi has been waiting for some years now to confront him on the question of the social aggregation of individual interests or utilities, just as he has confronted him behind the veil of ignorance. And the confrontation is exactly the same: what Harsanyi counters with is a version of the utilitarian principle.

If one listens to Harsanyi and others who have supported him, it is all over but the shouting. Utilitarianism, we are told, is an inevitable (i.e., logical) consequence of certain very simple and plausible axioms which are equally applicable to the problems of choice under conditions of uncertainty and social choice. These axioms, according to Harsanyi, are the essence of what he terms Bayesian rationality, and they lie at the very foundation of the remarkable work of Ramsey, von Neumann and Savage concerning the measurability of preference and subjective degrees of belief. In particular, they form the basis for, on the one hand, the von Neumann theory of choice under conditions of risk, and, on the other hand, the Ramsey–Savage theory of subjective probability. These constructions have assumed the status of a single *paradigm* (the Ramsey - Savage results for subjective probability really amount to nothing more than a very modest extension of the von Neumann results for 'objective' or canonical risk) – which many have chosen to ignore, but only a very few have chosen to take on, at least in any systematic fashion.[6]

Harsanyi's arguments cannot be dismissed lightly. Perhaps this accounts for why, even as early as *A Theory of Justice* (in contrast to the even earlier paper, 'Justice as Fairness') we find Rawls placing increased emphasis on that quite distinct line of justification, first articulated in his article, 'An Outline of a Decision Procedure for Ethics', in which appeal is made to the 'fit' between a theory of justice and the considered moral judgements of competent moral judges.[7] In the book, this line of reasoning is brought once again into prominence, as complementary to the analytic construction with its emphasis on what it would be rational to choose. The notion of 'fit' is there articulated in terms of the ideal of a theory which is gradually brought into reflective equilibrium with our more stable intuitions.[8] One can also suggest that the increasing emphasis which Rawls has placed on Kantian themes reflects once again his felt need to find some basis, other than a theory of rational 'economic' choice, in terms of which to critique the utilitarian position.

Leaving to one side, however, the effectiveness of these distinct lines of justification, one is bound to be bothered by the turn which the explication of rational choice itself has taken. One has only to recall the quite tortuous, and in the end quite unconvincing arguments of Bentham and Mill in support of the principle of utility, in order to wonder if Harsanyi could quite so easily accomplish what they were unable to do. What Harsanyi claims to have done is twice over to display a utilitarian kind of principle as a *theorem* of certain very simple axioms of rational choice (once in terms of axioms for individual rational choice under conditions of radical uncertainty, and once in terms of axioms for a reasonable social welfare function).[9] One can, of course, begin by pointing out that theoremhood is a logical relation: that for every proof via *modus ponens* there is the possibility of the converse argument against the axioms via *modus tollens*. Thus, if one is already disposed to doubt the validity of the principle of utility, one is likely to suppose that any conception of rational choice from which it can be logically derived is thereby suspect. That many have chosen not to construe the matter in this way is perhaps in the end testimony only to the entrenchment of the concept of rational choice in question: a conception to which we are principally indebted to the mathematical economists. There is, of course, much that could be said along this line. For my part, I find entrenchment as a criterion of validity for rules of behavior (as opposed to rules of linguistic behavior) about as subtle and telling as an old-fashioned appeal to tradition. I want to suggest, however, a somewhat different way of looking at this matter. Sometimes, entrenchment results from confusion, and I suggest that this is what has happened in the case of the axioms to which Harsanyi makes appeal. One of his axioms *seems* very reasonable but if we look at it very carefully, and if, in particular, we consider the kinds of argument which have been offered in support of it, it begins to look much less reasonable.

The critique that I shall propose turns out, I think, to cast a new light upon the dispute between Rawls and Harsanyi as it arises within both the context of models of individual choice under uncertainty and models of social choice. If I am correct, *both* Rawls and Harsanyi are wrong: for the axiom which I want to reject is one which is central to both of their constructions. This will also serve to explain, I think, why Rawls has had such a difficult time against Harsanyi. His acceptance of the axiom in question leaves him very little room to maneuver. This diagnosis is one which I have come to very reluctantly. Some two years ago I originally set out, upon reading Harsanyi's critique of Rawls in the *American Political Science Review*, to construct a defense of Rawls against Harsanyi.[10] The more I have explored this whole issue, however, the more I have become convinced that the only way to rebut

Harsanyi is at the price of giving up Rawls' constructions as well. In this respect I fear that what I have to propose here is something which no one will find pleasing. Moreover, my position has the distinct disadvantage of being only a negative position. I shall argue here that neither Rawls nor Harsanyi has managed to determine an appropriate principle of ordering for social alternatives: that once the suspect axiom is rejected, we face the familiar problem of underdetermination of a rule (or social choice function).[11]

2. THE SURE-THING AXIOM

The axiom to which I want to draw attention has been formulated in a variety of different ways, and given a variety of names. More often than not it is referred to as the 'sure-thing' axiom or principle. By way of approaching its formulation, we may think here of a typical decision problem (whether of the sort facing a single agent under conditions of uncertainty, or the sort involving the ranking of social alternatives) as representable in matrix form:

(1)

$$\begin{array}{ll} & B_1 \ldots B_k \ldots B_n \\ A_1 & u_{11} \ldots u_{1k} \ldots u_{1n} \\ A_i & u_{i1} \ldots u_{ik} \ldots u_{in} \\ A_m & u_{m1} \ldots u_{mk} \ldots u_{mn} \end{array}$$

Here the A_i's are understood as a finite set of basic alternatives from which choice is to be made (or over which a ranking is to be imposed).

If the decision problem is one of individual choice under uncertainty, the B_k's can be taken as a partition of events (that is, a set of mutually exclusive and exhaustive events), and the intersection of an alternative A_i and an event B_k as a (set of) consequence(s) to which there is associated a utility index u_{ik}. That is to say, on this interpretation, the agent has a utility function defined over the space of outcomes. The problem the agent faces, in this case, is that while his best choice, given reliable information concerning which event will take place, is fully determined by his utility function, he is presumed to be radically uncertain as to which event will take place. Thus each alternative constitutes a *vector* of possible utilities.

If the decision problem concerns the ordering of, or choice from, a set of social options, then the B_k's can be taken to designate the members of a set of n persons, and the cells in a given column specify the utilities for that person, of the alternatives A_1 through A_m. The decision problem then consists in giving a ranking of the A_i's (or determining an optimal set of A_i's) by means of some aggregation of the rankings of the n persons.[12]

In what is to follow, we shall employ the usual symbols for ordering relations: A_i R A_j means A_i is at least as good as A_j; A_i P A_j means A_i is preferred to A_j; and A_i I A_j means A_i is indifferent to A_j, where indifference is an equivalence relation. One other conceptual device that will prove useful is the following. Let B be the set of all events (or persons), and let B' be a proper subset of B. For each decision problem (A, B), we can construct another decision problem (A, B'), which results from restricting B to B', i.e., by eliminating from consideration all columns in $B - B'$. Correspondingly, let A_i/B' be the restriction of the option A_i (defined on B) to B'. With respect to any decision problem defined by such a restriction, we can ask, with respect to it, what is an appropriate ordering over the set of restricted alternatives.

The sure-thing axiom can now be formulated as:

(2) For all A_i and A_j such that for all B_k in $B - B'$, $u_{ik} = u_{jk}$:
$A_i \mathrm{R} A_j$ *iff* $A_i/B' \mathrm{R} A_j/B'$.

In short, the ordering of two alternatives is unaffected by the elimination of columns for which the two alternatives have the same utility.

It is possible to get by with an even weaker version of such an axiom, however, and one which harkens back to the axioms of von Neumann. I shall term this the mixture-version of the sure-thing principle, and it can be formulated as follows. Let A_i, A_j and A_k be three alternatives. Then A_i R A_j if and only if an even-chance mixture of A_i and A_k is at least as good as an even-chance mixture of A_j and A_k. That is, letting $(A_i \text{ ½ } A_j)$ be an even-chance mixture of A_i and A_j:

(3) $A_i \mathrm{R} A_j$ iff $(A_i \text{ ½ } A_k) \mathrm{R} (A_j \text{ ½ } A_k)$, for all A_k, A_i, and A_j.

Since this taken to hold for all A_k, we have, in particular:

(4) $A_i \mathrm{R} A_j$ iff $A_i \mathrm{R} (A_i \text{ ½ } A_j)$ and $(A_i \text{ ½ } A_j) \mathrm{R} A_j$.

and:

(5) $A_i \mathrm{I} A_j$ iff $(A_i \text{ ½ } A_j) \mathrm{I} A_i$ and $(A_i \text{ ½ } A_j) \mathrm{I} A_j$.

This last is a powerful combination of a convexity and an anti-convexity requirement. A mixture of two indifferent prospects is at least as good as either (convexity) and is no better than either (anti-convexity).[13]

By way of illustration, consider the following very simple choice problem, where the utilities in the cells have been scaled to the 0–1 interval, in the usual fashion.

(6)

	B_1	B_2	B_3	
A_1	1	0	1	
A_2	0	1	1	
A_3	x	y	z	$0 \leqslant x, y, z \leqslant 1.$

On the first formulation of the sure-thing axiom we have that A_1 R A_2 iff $(A_1/B_1 \cup B_2)$ R $(A_2/B_1 \cup B_2)$. On the second formulation we have that A_1 R A_2 iff $(A_1 \tfrac{1}{2} A_3)$ R $(A_2 \tfrac{1}{2} A_3)$.

3. THE PROBLEM WITH THE FIRST VERSION OF THE SURE-THING AXIOM

It needs remarking at the very outset that some of the confusion concerning the sure-thing axiom is terminological. The name, 'sure-thing' connotes a situation which is covered by a logically distinct, albeit related axiom. In the case of choice under conditions of uncertainty (as well as risk), this other axiom is usually referred to as a dominance requirement. A strong form of such a requirement is the following:

(7) If A_i and A_j are such that for all B_k in B, $u_{ik} \geqslant u_{jk}$, and for some B_k in B, $u_{ik} > u_{jk}$, then A_i P A_j. (Strong Dominance)

Use is often made also, of a somewhat weaker version:

(8) If A_i and A_j are such that for all B_k in B, $u_{ik} > u_{jk}$, then A_i P A_j. (Strict Dominance)

Conditions (7) and (8) have a natural interpretation as 'sure-thing' principles. The notion here is that, subject to an important proviso concerning the conditioning events being independent of the choice of an act (in some suitable sense of independence), A_i can be said to be a 'sure-thing' in the sense that, no matter what event occurs, one is always (in terms of the specified utilities) at least as well off having chosen A_i as one would be had one chosen A_j, and in at least one case one is better off.[14] Thus, in terms of outcomes, one cannot lose, as the saying goes, by choosing A_i over A_j. Conditions of this sort are intimately connected with the notion of rational choice as choice which maximizes one's preferences with respect to consequences (as represented by the utility assignments). With regard to (8), for example, to knowingly choose A_j over A_i, would be to choose with an expectation that one will end up with something less preferred than that which one would have ended up with, had one chosen A_i, regardless of what event takes place. And this, it is commonly argued, has to be irrational.

Turning to the case of social choice, the dominance conditions translate into the familiar Pareto conditions, which have traditionally been defended by reference to a plausible notion of group rationality, as social choice which does not leave unexploited the potential, in any situation, for mutual maximization or gain. Within the context of discussions of voting procedures (which may be viewed as a special case of social choice), the Pareto conditions correspond to the notion that unanimity for A_i over A_j (with or without abstentions) is decisive. More generally, their satisfaction in forced option situations signals a case in which there is no conflict of interest, but only mutuality of interest.

Now, the so-called 'sure-thing' principle speaks to a different consideration altogether. What it requires is that the ordering of alternatives be *independent* of the level of utility realizable under events for which this level is constant. That is to say, the ordering of A_i and A_j is a function just and only of those conditional orderings (individual orderings) under which choice one way or the other makes a utility difference. That such a requirement is far from self-evident is easily demonstrable by considering a third possible interpretation of our basic matrix representation of a decision problem. Let B now specify a set of different commodities, and A specify different possible bundles of varying amounts of these commodities. It is easy enough to construct examples in *this* sort of decision world where the sure-thing axiom is plausibly violated. Let A_1 be a bundle consisting of (a bottle of red wine; no white wine; a serving of steak) and A_2 be a bundle consisting of (no red wine; a bottle of white; a serving of steak). Clearly it casts no doubt on the rationality of an agent if he regards himself as indifferent between red and white wine, considered by themselves, but prefers the first to the second of the above two bundles. Yet this pattern of preferences violates the sure-thing axiom.

In the case of *conjunctive* bundles of commodities, we recognize the possibility that there will be interaction or complementarity between goods, and, hence, that the value of a bundle cannot necessarily be assumed to be a simple additive function of the value of its components. Some years ago Samuelson defended the imposition of such an independence requirement in the case of *disjunctive* prospects, on the grounds that in this case, by hypothesis, just and only one of a set of logically possible consequences will obtain: hence the sort of complementarity that arises with respect to conjunctive bundles cannot pose any problems.[15] But this is, as a defense of independence for disjunctive prospects, a *non-sequitur*. All his argument establishes is that one kind of complementarity cannot arise: it does nothing to show that there are no particular and distinct kinds of interaction which must be taken into account in evaluating uncertain prospects. Moreover, this argument works against the supposition that independence is a plausible condition on social choice: for in this case, as in the case of commodity bundles, we confront an instance in which the choice of a social policy leads to a conjunction of outcomes: one for each of the n persons involved.

I want to consider now a broad spectrum of cases in which both preference patterns for uncertain choices, and those for social choices, do in fact violate the independence axiom. In general terms, the independence axiom will be violated in all those cases in which evaluation of alternatives proceeds by reference not simply to considerations of (weighted or unweighted) averages, but also takes into account considerations of *dispersion*, at least in so far as the latter is measured in any of the familiar ways. We need not differentiate between uncertain and social choice here, for the same examples will do for both. Consider:

(9)

	B_1	B_2	B_3
A_1	0	1	1
A_2	$\frac{1}{3}$	$\frac{1}{3}$	1

By way of a relatively simple counter-example to independence, suppose that alternatives are evaluated in terms of the following formula:

$$(10) \qquad V(A_i) = M(A_i) - \tfrac{1}{2}n \sum_{k=1}^{n} (|M(A_i) - u_{ik}|).$$

This amounts to evaluation in terms of mean value discounted by one-half the average dispersion from the mean. In terms of this measure:

$$V(A_1/B_1 \cup B_2) = \tfrac{1}{2} - \tfrac{1}{4}(\tfrac{1}{2} + \tfrac{1}{2}) = \tfrac{1}{4}$$

$$V(A_2/B_1 \cup B_2) = \tfrac{1}{3}$$

$$V(A_1) = \tfrac{2}{3} - \tfrac{1}{6}(\tfrac{2}{3} + \tfrac{1}{3} + \tfrac{1}{3}) = \tfrac{4}{9}$$

$$V(A_2) = \tfrac{5}{9} - \tfrac{1}{6}(\tfrac{2}{9} + \tfrac{2}{9} + \tfrac{4}{9}) = \tfrac{11}{27}$$

By this formula we have that $(A_2/B_1 \cup B_2)\,\mathrm{P}\,(A_1/B_1 \cup B_2)$, but $A_1\ \mathrm{P}\,A_2$, in violation of the independence requirement.

Evaluation which incorporates a dispersion factor can be defended in the case of uncertainty choice on the following grounds. Given the usual assumption that the options are 'one-shot', there is no opportunity for approximation, over the long run, to the mean. Moreover, since by hypothesis the agent is uncertain as to the probability distribution over the events, he has no guarantee that the mean value corresponds to what his expectation would be, were he to possess information concerning the probability distribution. Caution might well suggest, then, that options of this type be evlauated in terms not simply of mean value, but mean value discounted by some measure of dispersion. In particular, the agent might suppose that a relevant consideration would be, that amount by which he might, in choosing a given option, fall short of its mean value. At the very least, it would be intelligible were he to prefer, when choosing between two options whose mean values are the same, the one with a smaller shortfall. But it is not unreasonable that he might go beyond this and regard increases in mean value as subject to a trade-off against improvements in shortfall. In particular, he might be willing to accept the certain prospect of an outcome with a fixed utility value to an uncertain prospect whose mean utility value is greater than that fixed value, but which involves a significant shortfall possibility. Thus, for example, he might well prefer the option of getting something with utility value 0.4, for certain, to the option of an uncertain prospect of outcomes with utility values 0 and 1, respectively. Now formula (10) incorporates such concerns. It is not, of course, the only way in which dispersion considerations could be factored into the evaluation procedure, and there are other ways in which dispersion itself can be characterized.[16] Moreover, it should be kept in mind here that we do not have to suppose that a rational person *must* take dispersion into account, in some manner or other. All that is necessary here is to suppose that concern for dispersion features is not irrational.[17]

With regard to the social-choice interpretation of the example given above, a similar line of reasoning can be constructed. The incorporation of a discount for dispersion would reflect the notion that, given full interpersonal comparability of utilities, an important consideration for social choice will be how equitably the benefits are distributed among the n persons involved. And, once again, an evaluative rule which factors in dispersion in the manner of (10) above, will relfect a concern that such equity considerations not enter simply by way of breaking ties between policies which promise the same mean value of benefits, but that they also be taken account of in a manner which permits trade-offs between mean value and dispersion (equity) considerations.

4. THE PROBLEM WITH THE SECOND OR MIXTURE VERSION OF THE SURE-THING AXIOM

So far I have directed my attention to the first of the versions of the sure-thing axiom. What about the other? Once again we confront a requirement concerning the relation between the ordering of certian entities, and the ordering of certain concatenations of them. The formal similarity between this version and the one previously discussed is obvious enough. We have here, once again, an independence, as distinct from a maximization, requirement. In this instance, however, the concatenation operation is interpreted as a probabilistic mixture, where this is to be understood in terms of some 'canonical' or 'standard' probability. Moreover, the axiom in this case applies univocally to both the uncertainty and the social choice problems (in contrast to the previous version, which required a different interpretaion of the concatenation operation for the two kinds of problems). We are here at the very heart of the von Neumann axiomatization of utility. Moreover, it could be argued that the axiom as applied to either the uncertainty problem or the social choice problem involves no (possibly suspect) extension of the von Neumann system, but only a making explicit what was already implicit in von Neumann's argument, namely, that in so far as various objects can be assigned a value, we can also say something about the value of canonical lotteries over such objects (regardless of the nature of these objects themselves – whether they be prizes, uncertain prospects, or social options, etc.)

We may bring this matter to a test in terms of the following simple example, which, once again, can be interpreted alternatively as a problem of social, or uncertainty, choice:

(12)

	B_1	B_2
A_1	1	0
A_2	0	1
A_3	$v(1\ \tfrac{1}{2}\ 0)$	$v(1\ \tfrac{1}{2}\ 0)$

Here we are to understand that A_3 yields under B_1 a lottery which itself yields something of value 1 with probability 0.5, and something of value 0 with probability 0.5; and that under B_2 it yields the same. With some plausibility we may assume that since A_3 yields $\nu(1\ \frac{1}{2}\ 0)$ constantly, any function taking utility vectors into an aggregate utility will have $\nu(A_3) = \nu(1\ \frac{1}{2}\ 0)$. Suppose now that the agent were to prefer A_3 to both A_1 and A_2. We shall return in a moment to consider what plausible stories might or might not be told in support of such a preference ordering. For the moment consider the implications of this ordering. By invocation of the mixture axiom, we can establish that if A_1 and A_2 are both dispreferred to A_3, then an even-chance mixture of A_1 and A_2 must also be dispreferred to A_3. But, whether we interpret B_1 and B_2 as (mutally exclusive and exhaustive) conditioning events, or as distinct persons to whom certain benefits are distributed, a plausible case can be made for the claim that an even-chance mixture of A_1 and A_2 is equivalent to A_3.

On the event interpretation of B, we can reason as follows. An even chance lottery over A_1 and A_2 can be interpreted as $(A_1/C_1; A_2/C_2)$, i.e., as a prospect of getting A_1 if C_1 occurs, and A_2 if C_2 occurs, where $P(C_i) = \frac{1}{2}, i = 1, 2$. But A_1 is itself a prospect of getting 1 if B_1 and 0 if B_2; and A_2 is a prospect of getting 0 if B_1 and 1 if B_2, i.e., $A_1 = (1/B_1; 0/B_2)$ and $A_2 = (0/B_1; 1/B_2)$. Substituting in, we have that an even chance lottery over A_1 and A_2 is simply: $((1/B_1;\ 0/B_2)/C_1;\ (0/B_1;\ 1/B_2)/C_2)$. Invoking the usual principles for the concatenation of *independent* events, we have that this is equivalent to: $(1/B_1\ \&\ C_1 \vee B_2\ \&\ C_2; 0/B_2\ \&\ C_1 \vee B_1\ \&\ C_2)$. By hypothesis, $P(C_i) = \frac{1}{2}$, from which we can conclude that no matter what the probabilities associated with B_1 and B_2 (recall that our assumption is that they are unknown), $P(B_1\ \&\ C_1 \vee B_2\ \&\ C_2) = P(B_1\ \&\ C_1) + P(B_2\ \&\ C_2) = P(B_1) \cdot P(C_1) + P(B_2) \cdot P(C_2) = \frac{1}{2}P(B_1) + \frac{1}{2}P(B_2) = \frac{1}{2}$, since $P(B_1) + P(B_2) = 1, B_1$ and B_2 being mutually exlusive and exhaustive events. Thus our even chance mixture of A_1 and A_2 is nothing more or less than an even chance at getting 1 or 0.

On the person interpretation of B_1 and B_2, the argument is even more straightforward. If B_1 expects to receive 1 if A_1 is implemented, and 0 if A_2 is implemented, then given that B_1 knows only that A_1 will be implemented with probability 0.5, etc., B_1 expects to receive 1 with probability 0.5 and 0 with probability 0.5. B_2's expectations are, of course, identical.

Dispreference for both A_1 and A_2 relative to A_3 thus leads to the unacceptable conclusion that an even chance mixture of A_1 and A_2 is worth less than what is, for all intents and purposes, something equivalent in value to it. But, of course, this quite unacceptable conclusion is obtained by appeal to the mixture condition. What must go, I suggest, is the mixture condition.

Dispreference for both A_1 and A_2 relative to A_3, while not required by rational choice, is surely permissible, both in the case of uncertainty choice, and the case of social choice.

In the case of uncertainty choice, it is the pattern of preference we expect from those who would, within limits, prefer known risks to unknown risks. A_1 and A_2 both offer the agent an uncertain prospect of getting either 1 or 0; A_3 offers the agent an even-chance prospect of getting 1 or 0. The reasoning by which we conclude that an even-chance mixture of A_1 and A_2 is equivalent in value to A_3 is compelling.[18] The claim that since both A_1 and A_2 are dispreferred to A_3, an even chance mixture of them cannot be as good as A_3, is unreasonable. To one who is averse to uncertainty, neither A_1 nor A_2 is worth very much, but a mixture is: for by mixing the two, the agent assures himself at least an even-chance at getting 1.

A parallel argument can be constructed in the social choice case. The pattern of preferences which offend the mixture condition are simply the preferences of all those who would agree with Hobbes, in his 13th and 14th Laws of Nature, concerning lot:

> But some things there be that can neither be divided nor enjoyed in common. Then the law of nature, which prescribes equity, requires that the entire right, or else – making the use alternate – the first possession, be determined by lot.

and:

> Of lots there be two sorts: arbitrary and natural. Arbitrary is that which is agreed on by the competitors ....[19]

Hobbes' prescriptions fit the present case like a glove. We are asked to consider how society shall choose when faced with just the alternatives A_1 and A_2. Both alternatives offend equity: but an even chance mixture of them does not: it gives each of B_1 and B_2 the same chance of receiving 1, and the same chance of ending up with 0.[20]

Those who find that these counter-examples do not weigh in with sufficient force to upset their intuitions about the mixture condition would do well to recall that we have the very best of precedents for the rejection of the mixture condition. von Neumann himself rejected it as a principle governing choice of strategies in a zero-sum game. The principal accomplishment of his theory of the zero-sum game was precisely to show that for a wide class of games, the rational solution consists of a mixture of strategies, no one of which it would be rational to play outright.

5. RECONSIDERATIONS

The formal work of the last twenty-five years makes it clear that the imposition of the sure-thing axiom, together with the usual ordering axioms, drastically limits the range of permissible rules. Luce and Raiffa summarize these results for the uncertainty case in Chapter 13 of *Games and Decisions*, and Sen has recently done the same thing for the social choice problem.[21] The general thrust of these theorems is to restrict the permissible rules to either some type of averaging principle or to some version of the minimax principle. Thus, for example, Deschamps and Gevers have shown that within the framework of full interpersonal comparability, and strong ordering requirements (full transitivity and independence of irrelevant alternatives), the strong Pareto condition, a strengthened form of Arrow's non-dictatorship condition, independence, and a modest assumption to the effect that sometimes the worst-off person gets his preference, suffice to narrow the choice to either some version of the average utility principle or the lexicographical version of the minimax principle.[22] The centrality of the independence condition can be appreciated when one notes that some years ago Arrow and Hurwicz offered a set of axioms for the uncertainty problem which, in contrast to the usual results, characterized a very broad set of rules: namely all those which are based on some weighting function over minima and maxima values. The independence axiom is conspicuously absent from this construction. If a version of it is appended, we are left with either just the minimax rule or the maximax rule.[23]

There is another disturbing feature of these constructions. We seek, by their means to articulate basic principles of rational individual and social choice. But the results which have been obtained within a framework constrained by the independence axiom are extremely brittle. What counts as a rational way to proceed has come to depend critically upon what sorts of informational and structural assumptions are imposed. For example, if it is assumed in the social choice case, that utility differences, but not levels, are comparable, then as d'Aspremont and Gevers have shown, the average utility principle is the unique solution. If, on the other hand, it is assumed that utility levels, but not differences, are comparable, then the lexicographical minimax principle is the unique solution. Again, assuming full interpersonal comparability, the addition of a continuity axiom suffices to establish the average utility principle as the unique solution.[24] It seems very strange that fundamental questions about the nature of rationality, individual or social, can turn so violently upon the informational and structural features of a given kind of decision problem.

On the whole, these are features of the constructions to which too little attention has been paid. The preoccupation has been with the question of how to choose between the averaging and the minimax alternatives. This a mistake. What Arrow showed many years ago was that if a very reasonable set of conditions were satisfied, the social choice rule must be dictatorial: there must be some person whose ordering of the social options determines the social ordering. With the introduction of some form or other of interpersonal comparison of utilities, this impass can be avoided – something which, ironically enough, was clearly anticipated in the work done in the early fifities on the uncertainty choice problem. But have we really escaped from the Arrow problem? The new results (as well as the uncertainty results first obtained in the early fifties) suggest that, at the level of comparable utilities, we face an Arrow-type problem once again. A reasonable set of conditions leads to the conclusion that the ordering will be 'dictated' by either the average person, or the ascending sequence of worst-off persons. All those rules which would try to strike some balance between considerations of average values and minima have been eliminated, and we are left with nothing by way of options except two extreme positions. The force of my analysis of the preceding sections should now be clear: it is the independence axiom which is responsible for this result.

From this perspective the dispute between Rawls and Harsanyi look somewhat different from the way in which it has usually been presented. I share Rawls' deep conviction that the principle of average utility is inadequate as a basic principle of social ordering. No matter what sort of 'rational construction' can be offered in its favor, it offends one's sense of equity. But within the framework which Rawls has implicitly accepted, this intuition can be cashed only by opting for an equally extreme position: the lexicographical minimax rule. This remains the case whether the argument proceeds in terms of what it would be rational to choose from behind the veil of ignorance, or in terms of reasonable conditions on a social welfare function. I do not mean to suggest here that no argument can be made for a very stringent equity requirement at the level of constitutional choice. But I think that argument must turn on something deeper than an appeal to the independence axiom, which is, after all, neither a moral requirement, nor a condition of maximization, but only a necessary condition for an additive representation.

Department of Philosophy, Washington University

NOTES

* This paper was prepared with support from the National Science Foundation, Grant SOC76-03915.

1 Rawls, J., *A Theory of Justice*, Cambridge, Mass: Belknap Press, 1971, in particular, Ch. 3. Harsanyi, J., 'Cardinal utility in welfare economics and in the theory of risk-taking', *J. Pol. Econ.* 61 (1953).

2 Cf. Arrow, K. J., *Social Choice and Individual Values*, 2nd Ed., Wiley, New York, 1963; Sen, A. K., *Collective Choice and Social Welfare*, Holden Day, San Francisco, 1970; Fleming, M., 'A cardinal concept of welfare', *Quart. J. Econ.*, 66 (1952); Harsanyi, J., 'Cardinal welfare, individualistic ethics, and interpersonal comparisons of utility', *J. Pol. Econ.* 68 (1955).

3 Perhaps the clearest of the presentations of formal results along these lines is: d'Aspremont, C. and Gevers, L., 'Equity and the informational basis of collective choice', *Rev. Econ. Studies* 46 (1977). Cf. also, Arrow, K. J., 'Extended sympathy and the possibility of social choice', *Amer. Econ. Rev.* 67 (1977).

4 Cf. Luce, R. D. and Raiffa, H., *Games and Decisions*, Wiley, New York, 1957, Ch. 13 for a very thorough review of these results.

5 Rawls, J., Appendix to Stanford Lectures (unpublished manuscript, 1978).

6 Two notable exceptions to this are: Levi, I., 'On indeterminate probabilities', *J. Phil.* 71 (1974); and Schick, F., 'Self-knowledge, uncertainty and choice', (forthcoming, *Brit. J. Phil. Sci.*, 1979)

7 Rawls, J. 'An outline of a decision procedure for ethics', *Phil. Rev.* 60 (1951).

8 Rawls, J., *A Theory of Justice*, Sections 9 and 87, in particular.

9 Cf. Harsanyi, J., 'Cardinal welfare, individualistic ethics and interpersonal comparisons of utility'. For more recent statements of his views, cf. his *Rational Behavior and Bargaining Equilibrium in Games and Social Situations*, Cambridge University Press, New York, 1977, Ch. 4; 'Bayesian Decision Theory and Utilitarian Ethics', *Amer. Econ. Rev. Papers and Proc.* 68 (1978). Many of his most important papers in this area are collected in J. Harsanyi *Essays on Ethics, Social Behavior, and Scientific Explanation*, D. Reidel, Dordrecht, Holland, 1976.

10 Harsanyi, J., 'Can the maximin principle serve as the basis for morality? A critique of John Rawls's theory', *Amer. Pol. Sci. Rev.* 59 (1975).

11 But those now beginning to work in this area can take solice in the consideration that there remains much to be done.

12 In this case, of course, the question of the degree of interpersonal comparison of utilities must be settled; whereas in the individual choice under conditions of uncertainty interpretation of the matrix, it is usual to assume that the agent has a utility function defined over the space of outcomes. All of this can be dealt with in terms of various informational constraints, or conditions specifying the uniqueness properties of the numerical representation of the problem. Cf. C. d'Aspremont and L. Gevers.

13 As we shall shortly see, the crucial assumption here is that of anti-convexity. The deep entrenchment of this condition in the literature on decision theory is curious in the light of the consideration that in the path-breaking work of von Neumann and Morgenstern on zero-sum games, anti-convexity is explicitly rejected. Cf. text below.

14 The question of what sort of independence of conditioning events from acts is

appropriate has recently been the subject of a vigorous debate. Cf. Gibbard, A. and Harper, W., 'Counterfactuals and two kinds of expected utility', in C. A. Hooker *et. al.* (eds.), *Foundations and Applications of Decision Theory*, Vol. I, D. Reidel, Dordrecht, Holland, 1978.

[15] Samuelson, P. A., 'Probability, utility, and the independence axiom', *Econometrica* 20 (1952).

[16] (10) amounts, in effect, to discounting the mean value by the *expected* shortfall, where that expectation is computed against the assumption that each of the entries is just as likely as any other. One could, of course, seek to use a standard deviation measure of dispersion, etc. But I am not convinced that the kinds of measures of dispersion which answer to the needs and interests of statistical inference are appropriate to the present problems.

[17] It is also possible to consider cases in which the agent might be oriented to, rather than averse to, uncertainty.

[18] Cf. Harsanyi, *Rational Behavior*, Ch. 3.

[19] Thomas Hobbes, *Leviathan*, Ch. 13.

[20] This is essentially the criticism which P. Diamond makes of Harsanyi's appeal to the mixture version of the sure-thing principle. Cf. his 'Cardinal welfare, individualistic ethics, and interpersonal comparisons of utility: a comment', *J. Pol. Econ.* 75 (1967).

[21] Sen, A. K., 'On weights and measures: informational constraints in social welfare analysis', *Econometrica* 45 (1977).

[22] Cf. Sen, 'On weights and measures . .' where this result is reported. The paper by R. Deschamps and L. Gevers is forthcoming in *J. Econ. Theory.*

[23] Cf. Luce and Raiffa, *Games and Decisions*, Ch. 13.

[24] Cf. Sen, 'On weights and measures . . .'.

RAYMOND DACEY

SOME IMPLICATIONS OF 'THEORY ABSORPTION' FOR ECONOMIC THEORY AND THE ECONOMICS OF INFORMATION

1. INTRODUCTION

This paper is in two parts. The first part recounts the implications of Morgenstern's concept of 'theory absorption' for three economic theories. First, the concept of the core, as it exists within the game-theoretical account of bilateral markets, is shown to lead to self-destroying predictions. Second, the traditional theory of consumer behavior is shown to be untestable in the face of absorption of alternative theories. Finally, the random walk theory of stock price behavior is shown to be immune to absorption, i.e., the theory is absorption proof. The second part of the paper presents the standard account of the economics of information and traces the implications of an inductive logic formalization of absorption upon that account. It is shown that the primary theorem of the standard account does not hold for the revised theory.

Morgenstern introduces the concept of 'theory absorption' as follows:

> Nature does not care – or so we assume – whether we penetrate her secrets and establish successful theories about her working and apply these theories successfully in predictions. In the social sciences, the matter is more complicated and in the following fact lies one of the fundamental differences between these two types of theories: the kind of economic theory that is *known* to the participants in the economy has an effect on the economy itself . . .
>
> However, the distribution of the kind of theory available, and the degree of its acceptance, will differ from one case to the other. This in turn will affect the working of the economy. There is thus a 'backcoupling' or 'feedback' between the theory and the object of the theory, an interrelation which is definitely lacking in the natural sciences.
>
> In this area are great methodological problems worthy of careful analysis. I believe that the study of the degree of 'theory absorption' by the members of the economy and the study of the above mentioned embedding relationship will make all of us more modest in judging how far we have penetrated into the economic problems [33, pp. 706–77].

The foregoing quotation expresses a refinement of ideas considered by Morgenstern throughout his life. Morgenstern began by questioning the capacity of economic theory to support predictions. In *Wirtschaftsprognose* [30], published in 1928, Morgenstern argues that economics is unlike any other science in that economic forecasts (i.e., predictions) can be understood by the agents within an economy, and that such predictions are self-falsifying.[1] In

J. C. Pitt (ed.), Philosophy in Economics, 111–136.

The Limits of Economics, a 1937 translation of the 1934 German edition, Morgenstern concerns himself with the application of economic theory to the problem of deciding upon an economic policy. He therein equates the application of economic theory with the construction of an outcome mapping for the decision maker, and details the limits of economics in that role [31, p. 1]. In 'Perfect Foresight and Economic Equilibrium', published (in German) in 1935, Morgenstern shows that perfect foresight and economic equilibrium are irreconcilable with one another [32], p. 174. Morgenstern returns to these points, and others, in the 1972 paper 'Thirteen Critical Points in Contemporary Economic Theory' [34]. The discussion of theory absorption, however, is presented separately in 1972 in 'Descriptive, Predictive and Normative Theory' [33]. The quotation above is an expression of an 'old' idea. However, there were no direct examinations of the implications of 'theory absorption' for economic theory until 1976. Part I of the paper recounts the implications of 'theory absorption' for economic theory developed in 1976.

Morgenstern describes 'theory absorption' but does not formalize the concept. Part II offers an inductive logic formalization of the concept, and presents a new implication of the concept for economic theory, specifically for the economics of information.

PART I

2. SOME IMPLICATIONS OF 'THEORY ABSORPTION' FOR ECONOMIC THEORY

2.1. *Absorption and the Core*

Morgenstern and Schwödiauer [35] show that the predictive power of the concept of the core, as it pertains to bilateral market games, is self-destroying. Specifically, they show

> ... that the core is a game-theoretic solution concept – i.e., a definition of rational behavior – which is not immune against 'theory absorption': It is an example of a theory of rational action the knowledge of which (on the part of the actors) destroys its predictive validity [35, p. 228, f.n. 11].

In this section we recount the argument advanced to support this claim. We begin with some definitions. A *bilateral market* consists of a finite set T of traders dealing in two different kinds of commodities. There are two types of *initial endowments,*

$$\langle a_i, 0\rangle \in \mathrm{Re}^2_+ \qquad \text{for } i \in M \subset T,$$

$$\langle 0, b_j\rangle \in \mathrm{Re}^2_+ \qquad \text{for } j \in N \subset T,$$

where the set M (= $\{1, 2, \ldots, m\}$) of sellers and the set N (= $\{m + 1, m + 2, \ldots, m + n\}$) of buyers constitute a partition of T. Let each trader possess a utility function U_i: $\mathrm{Re}^2_+ \rightarrow \mathrm{Re}$, and let the second commodity serve as an ideal standard of value, called 'money', which allows the traders to redistribute utility among themselves by means of side payments.[2]

A bilateral market game is defined by the characteristic function ν: $\mathscr{P}(T) \rightarrow \mathrm{Re}$, with

$$\nu(S) = \nu[(M \cap S) \cup (N \cap S)],$$

where $\nu(S)$ is the maximum utility or profit that a coalition S of traders can achieve by trading among themselves. Further restrictions are placed upon the bilateral market game (e.g., symmetry) by further specifying the characteristic function $\nu(S)$ as a function f(s, t), where s and t are the members of sellers and buyers, respectively, in the coalition S, i.e., $s = \#(M \cap S)$ and $t = \#(N \cap S)$.

We now come to the central concepts of the argument. An *imputation in* f is a point $\alpha \in \mathrm{Re}^{m+n}$ which satisfies

$$\alpha_i \geqq 0 \qquad \text{for all } i \in T, \text{ and}$$

$$\alpha(T) = f(m, n),$$

where $\alpha(S) = \Sigma_{i \in S}\, \alpha_i$, for all $S \subset T$. The set of all imputations in f is an $(m + n - 1)$-dimensional simplex; a particular imputation is a distribution of the maximum total profit among the m+n traders. Let α and β be some imputations in f; then α *dominates* β via coalition S if $\alpha_i > \beta_i$ for all $i \in S$ and $\alpha(S) \leqq f(s, t)$. The *core* of the game is the closed set G of all undominated imputations in f defined by

$$G = \{\alpha \mid \alpha(S) \geqq f(s, t) \quad \text{for all } S \subset T \text{ with}$$

$$s = \#(M \cap S) \quad \text{and} \quad t = \#(N \cap S)\}.$$

The core holds many attractions as a solution concept. The imputations associated with Walrasian parametric equilibrium prices are always elements of the core [45]. Furthermore, under certain conditions, the core contracts with increasing numbers of traders and coverages to the set of Walrasian equilibrium allocations [46]. Schotter [42] provides an extensive review of the concept of the core and its role in the game-theoritical analysis of general equilibrium.

Morgenstern and Schwödiauer present their argument as follows:

> There are, in general, imputations outside the core which are not dominated by any imputation in the core. Shapley's [44] symmetric market game with $f(s, t) = \min(s, t)$ ... may serve as an extreme example .... If the bargaining starts with an imputation ... in which some buyers are necessarily better off than in the core, and the buyers are aware that the procedure of free contracting and recontracting (i.e., competition among them) works by offering only temporary gains to some of them while making all worse off in the long run, they can be expected to conclude that it is profitable to stop the process of contracting and recontracting at some imputation outside the core which they are able to agree upon .... The above considerations may be summed up in the following, somewhat paradoxical, statement: If the traders are rational (in the sense of always striving for higher profits) and if they knew that the core was the only stable outcome of the bargaining process (and what the core looks like), then the core would not be stable! Thus, competition can only be expected to prevail if the behavior of the traders is characterized by a peculiar mixture of rationality, complete information about the opportunities the market offers, and short-sightedness [35, p. 228].

Morgenstern and Schwödiauer note that these difficulties can be overcome by replacing the core concept with the concept of a stable set solution as formalized in [36]. They also acknowledge that this shift induces a weakening of the notion of a solution: "... a stable set solution is not meant to predict a specific imputation (or price profile) or subset of imputations." ([35],p. 231). Thus, to avoid the self-destroying prediction of the theory of the core, we switch to the theory of the stable set solution, which does not admit of specific predictions. This switch appears to be a curious escape from self-defeating predictions only if we presume that prediction was the primary goal. The point at hand is that non-self-defeating predictions are impossible in certain market games. Compare this with Morgenstern's first methodological work [30] wherein he argues that forecasting (i.e., prediction) in economics differs from prediction in all other sciences in that economic forecasts of human behavior can be understood by the human agents, and will lead to "anticipations (i.e., expectations), which will render the original forecast false."[3]

2.2. *Theory Absorption and the Testability of the Theory of Consumer Behavior*

Dacey [5] provides an account of the effects of ambiguity upon the absorption and testability of alternative theories of consumer behavior. The traditional theory of consumer behavior posits that an individual behaves rationally by selecting commodities in amounts $x_1, x_2, \ldots, x_n$ so as to maximize utility subject to the budget constraint, i.e., by selecting $x_1, x_2, \ldots, x_n$ so as to

$$\text{maximize } u(x_1, \ldots, x_n)$$

$$\text{subject to } \sum_{i=1} x_i p_i = M.$$

Lancaster's 'new' account of consumer behavior [21, 22] posits that commodities are purchased for their endowed characteristics $z_1, z_2, \ldots, z_m$ and that the endowments are linear in the commodities, i.e., $z_j = b_{1j}x_1 + \ldots + b_{nj}x_n$. The Lancaster consumer behaves rationally by selecting $x_1, x_2, \ldots, x_n$ so as to

$$\text{maximize } u(z_1, z_2, \ldots, z_m)$$

$$\text{subject to } z_j = \sum_{i=1}^{n} b_{ij}x_i, \quad j = 1, \ldots, m$$

$$\text{and } \sum_{i=1}^{n} p_i x_i = M.$$

The traditional theory admits of two testable propositions, both concerning the Hicks–Slutsky term, defined to be

$$K_{ij} = \frac{\partial x_i}{\partial p_j} + x_j \frac{\partial x_i}{\partial M}.$$

The testable propositions are:

(1) $K_{ij} = K_{ji}$ for all i, j (symmetry), and

(2) $K_{ii} < 0$ for all i (negativity).

In the Lancaster theory (1) holds, but (2) does not hold, unconditionally. Dacey [5] shows that conducting a test of the traditional theory in an experimental economy depends upon the degree to which the two theories have been absorbed, the distribution of alternative inductive philosophies, and the ambiguity of the characteristics, none of which can be controlled by the experimenter. Therefore, the traditional theory is untestable in practice. The argument in support of this claim is presented in [5] and is not recountered here. The argument, following the comments of Morgenstern,[4] accounts for absorption as an inductive process, and formalizes the absorption of a theory as the acceptance of that theory via a particular cognitivist inductive logic.[5] The argument can be sketched as follows.

A cognitivist inductive logic is a pair $I = <P, U>$, where P is an inductive probability measure defined on an algebra of the sentences of a language and U is an (expected) epistemic utility function defined in terms of P. The measure P is based on the theory of inductive probability advanced by Hintikka [13–15]. The function U is the epistemic utility function introduced by Hintikka and Pietarinen [16]. The language upon which P is based is presumed to be a monadic, first-order language with k primitive predicates $R_1, R_2, \ldots, R_k$. Let $\lambda = \{R_1, \ldots, R_k\}$ and call the language L_λ. The (expected) epistemic utility function is specified as

$$U(g|e) = P(g|e) - P(g),$$

and measures the expected logical content of g relative to e. If a theory T is emplyed to introduce a new concept R_0, then T is a sentence in the expanded language $L_{\lambda \cup \{R_0\}}$. If T is conjoined with the evidence e, then U gives rise to two variants, namely

$$U_1(g|e \,\&\, T) = P(g|e \,\&\, T) - P(g)$$

and

$$U_2(g|e \,\&\, T) = P(g|e \,\&\, T) - P(g|T).$$

The adoption of either U_1 or U_2 to the exclusion of the other reflects the individual's philosophical position. U_1 seems fitted to the position of a (methodological) *instrumentalist*, whereas U_2 seems natural for a (scientific) *realist.*

The two inductive structures specified by U_1 and U_2 distinguish separate philosophies vis-a-vis the role of a *Weltanschauung* in the processing of potential additions to knowledge. A (methodological) instrumentalist views a theory T merely as a device of convience in organizing data, i.e., a theory is a mere instrument. A (scientific) realist takes a theory literally, adhering to the view that the theory (more or less accurately) describes reality.

The desired explication of 'theory absorption' can now be given. Let g be a theory (i.e., a generalization) and let $G(g)$ be the class of all those generalizations that compete with g. Then g is *absorbed given e* if and only if

$$U(g|e) = \max_{h \in G(g)} U(h|e),$$

and g is *absorbed given e* and T if and only if

$$U_i(g|e \,\&\, T) = \max_{h \epsilon G(g)} U_i(h|e \,\&\, T) \qquad (i = 1, 2)$$

There are thus three variants of absorption: absorption on the basis of factual evidence alone and both instrumentalist and realist absorption on the basis of factual evidence and new (theoretical) conceptual evidence. There are only two forms of inductive logic. Expanding upon the earlier characterization of an inductive logic,

$$I_1 = <P, U, U_1> \quad \text{is an instrumentalist logic of absorption,}$$

and

$$I_2 = <P, U, U_2> \quad \text{is a realist logic of absorption.}$$

It is therefore more accurate to speak of instrumentalist absorption and realist absorption.

A theory T in $L_{\lambda \cup \{R_0\}}$ is a *piecewise definition* of R_0 in terms of λ if and only if T logically implies a finite disjunction of explicit definitions of R_0 in terms of λ. Definition: a concept is inexact if and only if it is introduced by a *piecewise definition*. The sense of 'inexact' is specified as follows: if T is a *piecewise definition* of R_0 in terms of λ, then there are statements $C_1, \ldots, C_n$ (in $L_{\lambda \cup \{R_0\}}$ such that

$$T \Rightarrow (C_1 \Rightarrow Df_1) \,\&$$

$$T \Rightarrow (C_2 \Rightarrow Df_2) \,\&$$

$$\ldots$$

$$T \Rightarrow (C_n \Rightarrow Df_n),$$

where '$\Rightarrow$' denotes 'if ..., then ...,' and where each Df_i is a conjunction of explicit definitions of R_0 in terms of λ. The conditions $C_1, \ldots, C_n$ may each describe a separate context, which in turn specifies an exact definition for R_0. Thus the notion of piecewise definition explicates, at least in part, the concept of a contextual definition. The inexactness of a piecewise defined concept arises out of the alternative meanings the concept can possess in different circumstances. Piecewise defined concepts are, like explicitly defined concepts, both eliminable and noncreative. Thus piecewise definitions do not introduce new and unnecessary conceptual structures nor do they expand the underlying ontology of the language.

A generalization g is stable given T if and only if g is absorbed given e and g is absorbed given e and T. A theorem, based upon the results due to

Niiniluoto and Tuomela [39], asserts that there are two conditions such that the first will guarantee that a given generalization is stable for an instrumentalist, while it is not stable for a realist, and the second will guarantee that the same generalization is stable for a realist but not for an instrumentalist. Thus the condition which guarantees stability for a realist has no such effect on an instrumentalist, and conversely the condition which guarantees stability for an instrumentalist has no such effect on a realist. In an experiment involving an economy composed of both instrumentalists and realists, even if everyone begins with the same absorbed theory, the experimenter cannot guarantee the maintenance of the economic theory under test if inexact (i.e., piecewise defined) concepts are introduced or employed.

The preceding discussion of consumer theory is related to the analysis of theory absorption as follows: the matrix B is regarded as a complex theoretical predicate; the standard theory of consumer behavior is the complex generalization g_1 and Lancaster's theory is g_2; each individual possesses a (perhaps subjective) theory T which relates B to the predicates in λ; and the experimenter's instructions, i.e., the information catalyst, is the evidential statement e. The earlier results on absorption and maintenance of an economic theory can now be brought to bear in the specific context of testing the theory of consumer behavior.

If the agents in an experimental economy have each absorbed the traditional theory g_1, then, regardless of their inductive philosophies, if B is explicitly defined, then the traditional theory will be maintained. However, if B is introduced by a piecewise definition T, i.e., is inexactly specified or is contextual for an individual, then depending upon his inductive philosophy, maintenance of g_1 can be guaranteed for that agent by invoking the appropriate condition. If, as is more likely, the predicate B is introduced to separate individuals via piecewise definitions and these individuals adhere to different inductive philosophies, then the theorem shows that there is no single condition which guarantees that g_1 will be maintained.

Morgenstern's initial comments, as quoted in the Introduction, concerned the impossibility of performing a test of a theory in an economy involving agents with separate views. His concern can be extended to the dynamics of experimental testing over an economy involving agents who initially hold the same economic theory but adhere to separate inductive logics. For if inexact concepts are employed, there is no single set of conditions which guarantees maintenance of the theory under test.

2.3. *An Absorption-proof Theory*

Not all economic theories are self-falsifying or otherwise weakened when they are absorbed. Morgenstern offers the following example of an 'absorption-proof' theory. Definition: an economic theory is absorption-proof if, even when accepted by a majority of the participants in an economy, the theory cannot yield self-falsifying predictions. Then, the random-walk theory of the stock market is an absorption-proof theory. The random-walk theory of the stock market asserts

> that there cannot exist in that market a regularity that can be exploited upon discovery to their advantage (also allowing for costs of operations) by a sufficient number of actors on that market. No such (linear) regularities exist; this is equivalent with the discovery that price movements follow a random walk. This is a theorem with great predictive power .... Even if it becomes known and is applied it remains valid. This is due to its contents, which (are) indestructable by application. Other types of theories are destructable by application, as we have seen [33, p. 708].

Two aspects of this example are noteworthy. First, the random-walk theory of the stock market is a 'negative-theory' because it asserts that no exploitable regularities exist. This negativity suggests drawing a parallel with the fabled second law of thermodynamics, also a negative law, which asserts that no perpetual motion machine exists. The parallel is weak. A stronger parallel can be drawn by noting that the random-walk theory asserts that nature, here the (mindless) stock market, is playing a randomized (or mixed) strategy against the human agents. Thus, the parallel can be drawn between the absorption-proof random-walk theory of the (non-purposive) stock market and the theory of the mixed strategy, as optimal behavior, for a (purposive) human agent in a game situation. Secondly, the example suggests a conjecture: the concepts 'absorption-proof theory' and 'strategy-proof voting scheme' are interrelated so that a voting scheme is strategy-proof only if the theory of the voting scheme is absorption-proof.[6]

PART II

3. THE STANDARD ACCOUNT OF THE ECONOMICS OF INFORMATION

3.1. *Introduction*

This section reviews the standard account of the economics of information under the assumptions of costless information and costless decision, i.e., the standard problem of statistical decision. The adaptation of the standard account to the case of costly information (and costless decision) is discussed in Section 3.2. The review presented here draws heavily upon the work of Jacob Marschak, as presented in [27], [28], and especially [26].

The basic decision problem involves two decisions, the (logically prior) selection of an information source and the selection of a (Bayes) strategy of optimal responses to each of the possible messages from the information source. For example, the decision problem of what to do in response to different levels of oil pressure in an automobile engine involves, first, the selection of a detection system (traditionally, either an idiot light or an oil pressure gauge-type system) and, second, the selection of the optimal actions (specifically, 'stop the car' or 'do not stop the car') to be made in response to the possible signals from the system.[7]

3.2. *The Theory of Decision Making and an Example*

The basic decision problem is composed of a set Z of states of nature, selection from which is beyond the control of the decision maker, a set A of actions available to the decision maker, and a benefit function β with domain $A \times Z$ and range in the real numbers, Re, which measures the benefit to the decision maker of performing act a in state z. The decision maker ascribes initial (or 'prior') probabilities $P(z)$ to the states z in Z. The decision maker (initially) resolves the decision problem by selecting $a^0 \in A$ so as to maximize $\sum_Z P(z)\beta(a, z)$. The action a^0 is called the initial solution of the decision problem.

The decision maker can inquire about the state (future or present)[8] of the world. Such an 'inquiry' is a set Y, and each element y of Y is an 'answer' to the 'inquiry'. Similarly, Y is called an 'information system' and $y \in Y$ is a 'message' (or 'signal'). An information system can be uniquely characterized by a matrix η of the conditional probabilities (or 'likelihoods') $P(y/z)$ of the messages y given the states z.

The decision maker revises the inital resolution of the decision problem a^0 by selecting for each $y \in Y$ the optimal action $a^y \in A$ so that a^y maximizes $\sum_Z P(z/y)\beta(a, z)$.[9] The tuple $< a^{y_1}, a^{y_2}, \ldots, a^{y_n} >$, is called the *Bayes strategy*, and is the solution to the revised decision problem.

Consider a simple decision problem where $A = \{a_1, a_2\}$, $Z = \{z_1, z_2\}$, and the benefit and initial probability functions are given in the table below:

	$\beta(a, z)$		$P(z)$
	a_1	a_2	
z_1	30	10	0.5
z_2	8	30	0.5

The initial resolution of the problem is $a^0 = a_2$ since $\sum_Z P(z)\beta(a_1, z) = 9$ and $\sum_Z P(z)\beta(a_2, z) = 20$.

Now, begging the issue of the selection of an optimal information system, suppose that the decision maker has available, at no cost, the system Y with likelihood matrix η where $\eta = \begin{bmatrix} 0.8 & 0.2 \\ 0.4 & 0.6 \end{bmatrix}$. Simple computation yields the following probabilities:

	$P(z)$	$P(y/z)$		$P(z/y)$	
		y_1	y_2	y_1	y_2
z_1	0.5	0.8	0.2	0.4/0.6	0.1/0.4
z_2	0.5	0.4	0.6	0.2/0.6	0.3/0.4

Further, the revised resolution of the decision problem is the strategy $< a^{y_1}, a^{y_2} > = < a_1, a_2 >$ since

$$\sum_Z P(z/y_1)\beta(a_1, z) = \frac{13.6}{0.6} > \frac{10}{0.6} = \sum_Z P(z/y_1)\beta(a_2, z), \text{ and}$$

$$\sum_Z P(z/y_2)\beta(a_1, z) = \frac{4.4}{0.4} < \frac{10}{0.4} = \sum_Z P(z/y_2)\beta(a_2, z).$$

The foregoing example involves binary inquiry with two actions. This area is treated in general in Marschak [26, pp. 63–69].

3.3. *Properties of Information Systems*

An information system Y is *perfect* if and only if there is a 1–1 correspondence between Y and Z; i.e., Y is the complete partition $\{\{z_1\}, \{z_2\}, \ldots, \{z_n\}\}$. Otherwise Y is *imperfect.* an information system Y is *noiseless* if and only if all $P(y/z)$ are either 0 or 1. Otherwise Y is *noisy*. Clearly, if Y is perfect, then

Y is noiseless; however, the converse does not hold in general. Furthermore, if Y is a partition of Z, then Y is noiseless.

Y is a *null* information system if and only if all the rows of η are identical. A null system is denoted η^0. It is easy to show that, if a^0 is the initial (i.e., no-information) resolution of a decision problem, then the resolution of the decision problem revised on the basis of a null system is the Bayes strategy $< a^0, a^0, \ldots, a^0 >$. That is, a null information system will not alter the decision maker's initial decision.

An information system Y' is *a garbling* of system Y if and only if there is a markov matrix G of order $m \times n'$ such that $\eta' = \eta G$. The elements of G are the conditional probabilities $P(y'/y)$, and the garbling Y' is characterized by the likelihood matrix $\eta' = [\sum_Y P(y'/y)\, P(y/z)]$.[10]

An information system Y is *finer than* system Y' if and only if Y' is a noiseless garbling of Y. That is, G is reduced to a many-to-one mapping, g, from Y (of size n) to Y' (of size n') with $P(y'/y) = \{{1 \atop 0}\}$ if $y' \{{= \atop \neq}\}\ g(y)$. Clearly, $n' \leqslant n$. If Y and Y' are themselves noiseless, then Y is finer than Y' if and only if Y is a subpartition of Y'.

Finally an information system Y is *useless* if and only if the same action is optimal for all messages; that is, $a^y = a_k$, say, for all y. Clearly, all null information systems are useless, but the converse is not, in general, true.

3.4. *Comparative Informativeness*

The decision maker faces the problem of selecting an optimal information system. To solve that problem the decision maker must be able to totally order the set of alternative information systems. To do so the decision maker assigns a value to each information system, determined for the costless case, as follows:

$$V(Y) = V(\eta) = \sum_Y P(y) \max_a \sum_Z P(z/y)\beta(a, z)$$

$$= \sum_Y P(y) \sum_Z P(z/y)\beta(a^y, z).$$

For the foregoing example, we have $V(Y) = 0.6\left(\frac{13.6}{0.6}\right) + 0.4\left(\frac{10}{0.4}\right) = 23.6$. Note that the Bayes strategy can be viewed as a lottery ticket obtained from the information system (at no cost), paying $\sum_Z P(z/y)\beta(a^y, z)$ with probability $P(y)$, for each $y \in Y$. Then $V(Y)$ is simply the expected benefit from the lottery ticket, i.e., from the information system. The set of all (cost-free) information systems on Z can be ordered by V, because it is a real valued mapping.

Recall that $V(Y)$ depends upon the specification of the initial probabilities and the benefit function. Following Blackwell [4], an alternative ordering is defined as: η *is more informative than* η' if and only if $V(\eta) \geqq V(\eta')$ for all specifications of $P(z)$ and β. Denote that η is more informative than η' by $\eta > \eta'$. The qualifier 'for all specifications of $P(z)$ and β' renders $>$ a partial ordering.

We now consider the relationships that exist between the various properties of information systems and the partial ordering $>$.

THEOREM. If η^{∞} is a perfect information system and η^0 is a null system then $V(\eta^{\infty}) > V(\eta) > V(\eta^0)$ for all η on Z. Furthermore, $V(\eta^{\infty})$ is finite [26, p. 55].

Thus, the perfect information system and the null information system are the maximal and minimal elements, respectively, in the lattice induced on the set by the partial ordering $>$ of information systems on Z.

As was noted above, not all useless systems are null. However, all useless systems are equivalent to the null system with respect to the ordering $>$. Formally,

THEOREM. $V(\eta) = V(\eta^0)$ if and only if η is useless [26, p. 52].

Intuition suggests that a garbling of a message cannot be more valuable than the original message, because the decision maker will react only by chance to a garbling of the message y with the appropriate action a^y. Formally, we have

THEOREM. If η, η' and G are Markov matrices with $\eta' = \eta G$, then $\eta > \eta'$ [26, p. 54].

Similarly, intuition suggests that a refinement of an information system is at least as informative as the original system and, therefore, given costless information, at least as valuable as the original. Formally, *the primary theorem of the economics of (costless) information* is

THEOREM. If η is finer than η', then $\eta > \eta'$ [26, p. 56].

The foregoing theorems relate garbling and fineness respectively, to the ordering $>$. They are weak in that each provides sufficient, but not necessary, conditions for $>$ to hold between two information systems. The following two theorems strengthen the foregoing by providing necessary and sufficient conditions:

THEOREM. $\eta > \eta'$ if and only if $\eta' = \eta G$ for some Markov matrix G,

and

THEOREM. If η and η' are noiseless, then $\eta > \eta'$ if and only if η is finer than η' [26, p. 56].

The first is known as Blackwell's theorem [4], and asserts, unconditionally, that η is more informative than η' if and only if η' can be formulated as a garbling of η. The second theorem asserts, conditionally, that for noiseless information systems, η is more informative than η' if and only if η is a refinement of η'. The connection between the theorems is as follows: suppose the garbling G is noiseless; then G reduces to a mapping g: $Y \to Y'$ with $P(y'/y) = \{{1 \atop 0}\}$ if y' $\{{= \atop \neq}\}$ $g(y)$, for all y, y'. That is, G is reduced, in the noiseless case, to a many-to-one mapping g from Y, of size n, to Y', of size n', where clearly $n' \leqslant n$. Thus, Y is finer than Y' [26, pp. 55–56].

3.5. *Costly Information*

The decision problem, given costless information, is the statistical decision problem and is resolved in two parts as discussed above. If information is costly, then the selection of an optimal information system must account for that cost. Inasmuch as comparative informativeness was defined in terms of the (economic) valuation function V, accounting for the cost of a system requires minor adjustments.

Let $K(Y)$ denote the cost of installing and operating the system Y.[11] For example, K (Wall Street Journal) is the subscription price, while the cost of an earth resource detection satellite is somewhat higher. Consider the benefit function β. Following von Neumann–Morgenstern [36, pp. 49–50] and Savage [40, pp. 13–17], we write β as a composition of an outcome (or consequence) mapping ρ and a utility function μ, i.e., $\beta = \mu \circ \rho$.[12] The outcome mapping has as its domain $A \times Z$ and as its image a set 0; the utility function has as its domain 0, and as its range the real numbers, Re. The set 0 may be a subset of the real numbers, but is not required by the formulation of the standard account to be a subset of the reals. As Savage notes: "A consequence (i.e., an outcome) is anything that may happen to the person." [40, p. 13] If the outcomes are numerical and the cost of the system is $K(Y)$, then we can define the net benefit of performing act a in state z given system Y to be $\mu(\rho(a, z) - K(Y))$. To simplify the following analysis we presume that (i) $0 \subset$ Re, (ii) net benefit can be written as $\mu(\rho(a, z) - K(Y))$, (iii) the cost function K is independent of a and z, and (iv) the utility function is continuous and strictly increasing but not necessarily linear. Of the last assumption, note that the net benefit

$$\begin{aligned}\mu(\rho(a, z) - K(Y)) &= \mu(\rho(a, z)) - \mu(K(Y)) \\ &= \beta(a, z) - \nu(Y)\end{aligned}$$

if and only if the utility function μ is linear in its argument [28, pp. 84–85]. Here, $\nu(Y) = \mu(K(Y))$ is interpreted as the cost of Y in utility units. Again, the following analysis does *not* presuppose that μ is linear.

There are two alternate concepts of the value of a costly information system. The cost of no information can be taken to be 0 by suitable choice of the unit for the outcome mapping ρ [28, p. 85]. Then the maximum net expected utility with no information is

$$\max_a \sum_Z P(z)\mu(\rho(a, z)) = \max_a \sum_Z P(z)\beta(a, z) = \sum_Z P(z)\beta(a^0, z).$$

The *monetary value of an information system* Y, denoted $MV(Y)$, is defined to be the solution to the equation

$$\sum_Y P(y) \max_a \sum_Z P(z/y)\mu[\rho(a, z) - MV(Y)] = \max_a \sum_Z P(z)\mu(\rho(a, z)).$$

Assumptions (i)–(iv) guarantee the existence of a unique solution to the above equation. [28, p. 86]

An alternative concept, *the net value of an information system* Y, denoted $NV(Y)$ is defined as follows:

$$NV(Y) = \sum_Y P(y) \max_a \sum_Z P(z/y)\mu(\rho(a, z)) - \max_a \sum_Z P(z)\mu(\rho(a, z)).$$

The net value function $NV(Y)$ is related to the valuation function $V(Y)$ considered above as follows:

$$\begin{aligned}NV(Y) &= V(Y) - \max \sum_Z P(z)\mu(\rho(a, z)) \\ &= V(Y) - V(Y^0),\end{aligned}$$

where Y^0 is the null information system. Thus, NV and V induce the same ordering on the set of all information systems for Z.

The valuation functions' MV and NV are related in two ways.[13] First, if the utility function μ is linear, then NV and MV are identical. Second, the theorem on fineness stated for V above holds for MV, i.e., formally,

THEOREM. If Y and Y' are distinct information systems for Z, then if Y is finer than Y', then $MV(Y) > MV(Y')$, for all specifications of $P(z)$, ρ and μ [28, p. 86].

Thus the primary theorem in the standard account of decision making and the economics of information holds for both costless information (the statistical decision problem) and costly information. We now show that this theorem does not hold if absorption in the sense of Morgenstern is permitted.

4. THE IMPLICATIONS OF 'THEORY ABSORPTION' FOR THE STANDARD ACCOUNT OF DECISION MAKING AND THE ECONOMICS OF INFORMATION

4.1. *The Standard Account of Outcome Mapping Selection*

The point at hand is to show that, if the decision maker is capable of absorbing a theory of relevance to his own behavior, then the primary theorem of the economics of information fails to hold in general. To do so we must formalize Morgenstern's (unformalized) notion of absorption and, consequently, alter the standard account of the economics of information.

The concept of information employed in the standard account of the economics of information pertains to the content of singular messages, i.e., messages expressed by singular propositions. Morgenstern's concept of absorption pertains to generalizations, i.e., whole theories, especially economic theories. The application of an economic theory in the selection of an economic policy involves the prediction of the consequences of that a policy in each of the possible (future) states of nature. As Morgenstern puts it:

> All those who are concerned with the introduction of proposed measures of economic policy have to consider what the effects will be .... The appraisal of the effects depends, firstly, on a judgement as to the actual condition of the economic system already prevailing, that is, on a fairly exact knowledge of facts. It depends, secondly, on the view as to how the course of economic events will be changed as a consequence of the acts of intervention ... [31, p. 1].

This view can be formalized by introducing a mapping I with Y as its domain and a set of outcome mappings as its range. The construction of the set of outcome mappings is trivial if the outcome set O is given; the construction in non-trivial if the set O is not specified. The selection of I is a problem in (congnitivist) inductive logic, and a number of alternative methods of resolving the selection problem have been advanced.[14]

The standard account of decision making does not formally address the problem of selecting an outcome mapping, primarily because of the way in which the concept of an outcome is introduced into the formal structures of the various theories. von Neumann and Morgenstern presume that the outcomes of the various players' actions are specified as part of the game [36, p. 50]. The decision problem faced by the individual decision maker is formalized as a two-person game between the decision maker and nature, the latter a non-human player restricted to randomized behavior. For von Neumann and Morgenstern, there is no formal problem of selecting an outcome mapping – it is specified as a part of the decision problem.

Savage [40] formalizes an act as a function from a set of states of nature to a set of consequences. There is a flaw in Savage's formalism – it does not admit of an orderly interpretation. The flaw is noted by Jeffry as follows:

> For Savage, choice of an 'act' is choice of a scheme which associates a definite consequence with each of the infinity of possible states of nature. If these are acts, then only God could know what act is being performed. After performance, the human agent may learn what consequence his act associated with the actual state of nature, but neither before nor after its performance can he be expected to know what consequences the act associates with the rest of the possible states [18, p. 74].

To remedy this and other flaws, Balch and Fishburn [2] have advanced a theory of decision making where the occurence of a state is dependent upon the action selected (and performed) by the decision maker. The Balch–Fishburn theory resolves the problem discussed above by Jeffrey attendant to the Savage theory and further problems attendant to the Krantz–Luce theory of decision making given act dependent states.[15] The Balch–Fishburn theory takes as primitive the set F of acts presumed available for choice, and introduces, for each act f, the (mutually exclusive and collectively exhaustive) set S_f of logically possible 'states-of-the-world-as-conditioned-by f'. Let E'_f be a Boolean algebra of subsets of S_f, devoid of the empty set. Then $A \in E'_f$ is an f-conditional event. The decision maker is then presumed to have a coherent preference order over the set of act–event pairs $\langle f, A \rangle$, and

> ... is to think of $\langle f, A \rangle$ as that circumstance, perhaps uncertain still, which is conditioned by choice of f and suppositioned on the occurence of A. That is, he is to imagine what might happen if indeed the true f-conditional state were in A [2, p. 61].

Thus far Balch and Fishburn have engaged only in formalities. If an individual were to apply their theory, he would then have to determine what would happen when indeed the true f-conditioned state is in A, for each act f and each event A. That is, the decision maker must employ an outcome mapping that assigns to each pair $< f, \{s\} >$, when $s \in A \in E'_f$, the outcome realized if act f is performed and event A includes the true state of nature. As in the Savage formulation of decision making, the problem of the selection of an outcome mapping is not treated overtly but rather is implicity presumed to be resolved.

An alternate (formal) move is made by Jeffrey [12], wherein the decision maker is presumed to have a coherent preference ordering on a Boolean algebra of the set of all holistic states, which "... specify what act the agent performs, along with everything else one usually takes it to specify" [18, p. 75]. The objects of preference are then propositions, i.e., sets of states. This formal move like the one employed by Balch and Fishburn, simply presupposes the solution of the outcome mapping selection problem. To apply Jeffrey's theory, the decision maker must determine the proposition that describes the states of the world given the performance of each possible action in each of the initially possible states of the world.

Savage characterizes the construction of an outcome mapping as a fairly trivial exercise. He does so via the (now famous) example of the omelet problem.

Your wife has just broken five good eggs into a bowl when you come in and volunteer to finish making the omelet. A sixth egg, which for some reason must either be used for the omelet or wasted altogether, lies unbroken beside the bowl. You must decide what to do with this unbroken egg. Perhaps it is not too great an over-simplification to say that you must decide among three acts only, namely, to break it into the bowl containing the other five, to break it into a saucer for inspection, or to throw it away without inspection. Depending on the state of the egg, each of these three acts will have some consequences of concern to you, say that indicated by Table 1.

Table 1. An Example Illustrating Acts, States and Consequences

	State	
Act	Good	Bad
Break into bowl	Six-egg omelet	No omelet and five good eggs destroyed
Break into saucer	Six-egg omelet, and a saucer to wash	Five-egg omelet, and a saucer to wash
Throw away	Five-egg omelet, and one good egg destroyed	Five-egg omelet

[40, pp. 13–14]

Savage, after formalizing the concept of an act as a mapping from the set of states to the set of consequences, makes the following comment:

> The argument might be raised that the formal description of decision that has thus been erected seems inadequate because a person may not know the consequences of the acts open to him in each state of the world. He might be so ignorant, for example, as not to be sure whether one rotten egg will spoil a six-egg omelet. But in that case nothing could be simpler than to admit that there are four states in the world. Corresponding to the two states of the egg and the two conceivable answers to the culinary question whether one bad egg will spoil a six-omelet. It seems to me that this solution works in the greatest generality, though a throughgoing analysis might not be trivial. A reader interested in the technicalities of this point ... will find an extensive discussion of similar problem in Chapter II of [30], where von Neumann and Morgenstern discuss the reduction of a general game to its reduced form [40, p. 15].

The omelet example is deceptively simple; any individual who did not know the consequences for the omelet problem could seemingly resolve his 'knowledge' problem in the way that Savage proposes. Consider an alternate imaginary problem. The decision maker must select the protective (i.e., anti-radiation) wrapping for an on-board experimental apparatus on a probe to Jupiter. The acts available to the decision maker consists of the alternative materials and the thickness with which they are applied (e.g., 2 cm gold foil, 3 cm gold foil, etc.). The states of the world are the levels of radiation encountered along the path that the probe takes through the radiation belts of Jupiter. The consequences of the decision maker's acts in the various states of the world cannot be ferreted out via Savage's technique but must be predicted via physical theory. Clearly, the decision maker could be aware that he did not 'possess' such a theory and thereby be aware that he did not know the consequences of his acts. Similarly, the decision maker may possess a theory of the required type, but may be challenged by a second party on the correctness of the theory.

4.2. *The Economics of Information Given Cognitivist Absorption*

The present (revised) account acknowledges that the decision maker faces a two-part problem, i.e., the resolution of the decision problem and the selection of an optimal information system. The present account posits that each signal y from an information system yields revised state probabilities $P(z/y)$, as per the standard account but also that each signal y induces the absorption (i.e., acceptance) of a particular outcome mapping ρ_y. Let ρ_o denote the outcome mapping absorbed before the receipt of the signal y.

Then, the initial decision problem is resolved by selecting $a^0 \in A$ so as to maximize $\sum_Z P(z)\mu(\rho_O(a, z))$. The revised problem is resolved by selecting a strategy $< a^{y_1}, \ldots, a^{y_n} >$ such that a^{y_t} maximizes $\sum_Z P(z/y_t)\mu(\rho_{y_t}(a, z))$. The selection of an optimal information system is based on V if the system is costless and on MV if it is costly. Note that the selection of the (optimal) information system Y is an economic problem involving the act set A, the state set Z, and the inductive logic I but that the selection of the (optimal) inductive logic I is *not* an economic problem and does not depend upon A and Z.[16]

The primary theorem of the standard account of the economics of information asserts that, if Y is finer than Y', then $V(Y) > V(Y')$, if Y, Y' are costless, and $MV(Y) > MV(Y')$, if Y, Y' are costly. Given the cognitivist absorption of an outcome mapping ρ_y given the signal y, $V(Y)$ is

$$V(Y) = \sum_Y P(y) \max_a \sum_Z P(z/y)\mu(\rho_y(a, z)) - \max_a \sum_Z P(z)\mu(\rho_O(a, z)),$$

and, similarly, $MV(Y)$ is determined by solving the equation

$$\sum_Y P(y) \max_a \sum_Z P(z/y)\mu(\rho_y(a, z) - MV(Y)) = \max_a \sum_Z P(z)\mu(\rho_O(a, z)).$$

The claim is: there exist information systems Y and Y' such that Y is finer than Y' but $V(Y) < V(Y')$ and $MV(Y) < MV(Y')$, i.e., the primary theorem of the standard account of the economics of information does not hold given cognitivist absorption. An example is offered to substantiate the claim. In the example, the utility function μ is linear and thus guarantees that V and MV induce the same order on the set of information systems.

4.3. *A Counterexample to the Primary Theorem of the Standard Account*

The claim was made above that the primary theorem of the economics of information does not hold given theory absorption. To substantiate this claim, it is sufficient to provide an example of a decision problem involving two information systems Y and Y' where Y is finer than Y' but $MV(Y) < MV(Y')$. The following example is offered to substantiate the claim.

Consider a decision problem with $A = \{a_1, a_2\}$, $Z = \{z_1, z_2\}$, and initial outcome mapping $\rho_O = \begin{pmatrix} 20 & 10 \\ 10 & 30 \end{pmatrix}$.[17] Let the decision maker initially hold the probabilities $P(z_1) = 0.6$, $P(z_2) = 0.4$, and let the decision maker's utility function be the simple linear function (o) = o for each outcome $o \in O$. Then the initial resolution of the problem is a_2, and $\max_a \sum_Z P(z)\mu(\rho(a, z)) = \max \{16, 18\} = 18$.

Now, let Y be characterized by the likelihood matrix $\eta = \begin{pmatrix} 0.6 & 0.3 & 0.1 \\ 0.2 & 0.1 & 0.7 \end{pmatrix}$, and let $\rho_{y_1} = \rho_{y_2} = \rho_{y_3} = \begin{pmatrix} 20 & 15 \\ 10 & 20 \end{pmatrix}$. Then the resolution of the decision problem revised on the basis of Y is the strategy $< a_1, a_1, a_2 >$ and

$$\sum_Y P(y) \max_a \sum_Z P(z/y)\mu(\rho_y(a, z) - MV(Y)] = 18.5 - MV(Y).$$

Therefore, $MV(Y) = 0.5$.

Consider a noiseless garbling Y' of Y obtained via $G = \begin{pmatrix} 10 \\ 10 \\ 01 \end{pmatrix}$. Then $\eta' = \begin{pmatrix} 0.9 & 0.1 \\ 0.3 & 0.7 \end{pmatrix}$. Furthermore, let $\rho_{y'_1} = \rho_{y'_2} = \rho_O$. Then the resolution of the decision problem revised on the basis of Y' is the strategy $< a_1, a_2 >$, and

$$\sum_{Y'} P(y') \max_a \sum_Z P(z/y')\mu[\rho_{y'}(a, z) - MV(Y')] = 21 - MV(Y').$$

Therefore, $MV(Y') = 3.0$.

The foregoing example substantiates the claim that the primary theorem of the standard account of the economics of information does not hold if the decision maker can absorb predictive economic theories. One caveat must be made. If both Y and Y' are noiseless, then they are partitions of Z, and hence a message $y \in Y$ or $y' \in Y'$ would introduce nothing 'new'. That is, if the decision maker has initially accepted the outcome mapping ρ_O: $A \times Z \to 0$, then the receipt of a noiseless signal, either from Y or Y', which is simply a subset of Z ought not, as intuition suggests, induce the decision maker to absorb a new outcome mapping. Thus the claim made earlier should be restricted to noisy systems Y and Y'.

It is not surprising that the primary theorem of the standard account fails to hold in the revised absorption account. Recall that an inductive logic I selects outcome mappings, not on the basis of the economics of the decision problem at hand but rather on epistemic grounds. Thus, an inductive logic I may select, given a signal y, an outcome mapping ρ_y very unlike the initial outcome mapping ρ_0. Therefore, the construction of a counter-example is a trivial exercise. The triviality of the exercise is due to the separation of I from the basic (economic) problem. The separation is based upon our intuitions about inductive logic. Hempel's previously cited remark captures those intuitions.

5. CONCLUSIONS

Morgenstern's principle concern is with the self-falsifying nature of public predictions and economic theories. A second concern, stressed in the analyses of consumer behavior and the economics of information, is with the (methodologically) disruptive aspects of absorbable theories. If (economic) theories are employed as outcome mappings in decision making situations and if 'absorption' is explicated as inductive acceptance, then theory absorption disrupts the standard accounts of consumer behavior and the economics of information. The foregoing analyses of consumer behavior and the economics of information do not claim that absorption of these theories would lead to their self-falsification. Rather, the foregoing analyses claim that the acceptance of outcome mappings by the agents in a decision making situation (1) render consumer theory untestable, and (2) the principle theorem of the economics of information false.

The foregoing analyses lead to a problem in the foundations of inference. The problem is centered upon the process for selecting an outcome mapping. The foregoing analyses equate absorption with (inductive) acceptance. There are two basic positions on inductive inference. The cognitivist position maintains that rules of acceptance and rejection of hypotheses are indespensable and that knowledge consists of accepted hypotheses. The behavioralist position maintains that inductive inference consists of the assignment of probabilities to hypotheses and that these probability assignments are to be employed in the resolution of practical decision problems [37, p. 148]. The two positions are incompatible. The problem is captured in the following claim: the behavioralist position presupposes the cognitivist position. An argument in support of the claim would go beyond the scope of the present paper. The following is an argument sketch.

A decision maker resolves a practical (decision) problem by selecting an act a so as to maximize his own expected utility $EU(a) = \sum_o \sum_s P(o \mid s \,\&\, a)$ $P(s \mid a)\, U(o)$, where: act a performed in state s induces outcome o with probability $P(o \mid s \,\&\, a)$; act a induces state s with probability $P(s \mid a)$; and $U(o)$ is the decision maker's utility of outcome o. If the acts and states are independent, then $P(s \mid a) = P(s)$ for all a, s.

The probability $P(o \mid s \,\&\, a)$ accounts for the outcome mapping of relevance to the decision problem. If the outcome mapping is deterministic, i.e., if each act–state pair $< a^*, s^* >$ uniquely determine an outcome o^*, then $P(o^* \mid s^* \,\&\, a^*) = 1$ and $P(o \mid s^* \,\&\, a^*) = o$ for all $o \neq o^*$. If, however, the outcome mapping is statistical, then $P(o \mid s \,\&\, a)$ will take on non-integer values. The distinction between deterministic and statistical outcome mappings parallels the distinction between deterministic and statistical laws made in the Hempelian accounts of deductive-nomological and statistical explanation, respectively.

The crux of the argument is as follows: whether the outcome mapping is deterministic or statistical, it is expressed by a synthetic statement (or statements). Synthetic statements can be employed in the resolution of decision problems only if they are accepted as knowledge [20, pp. 113, 162–163; 38, p. 142]. Either the acceptance of synthetic statements is logically prior to the resolution of practical decision problems, or practical decision problems can not be resolved. That is, either the behavioralist account of inductive inference presupposes the cognitivist account, or the behavioralist account is incomplete.

The foregoing sketch does not constitute a complete argument for the claim. The claim, however, suggests that the concept of theory absorption leads to major problems in the methodological foundations of inference as well as in the foundations of the social sciences.

University of Oklahoma

NOTES

1 See [30, pp. 92f], and [25, p. 314]. An English translation of [30] is not available. However, the review article by Marget [25] is very informative. Morgenstern's claim is refuted by Grunberg and Modigliani [9].

2 A comment on notation: Re will denote the real numbers throughout the paper. Re_+^2 is the Cartesian product of the set of positive reals with itself, i.e., $\mathrm{Re}_+^2 = \{<x, y> | x, y \in \mathrm{Re} \text{ and } x, y \geqq 0\}$. The definitions given here are taken directly from [35].

3 This is Marget's translation of Morgenstern's comment. See [30, pp. 92–106], and [25, pp. 314 and p. 326, f.n. 39].

4 Specifically, [31, p. 1] and [33, pp. 706–707], the latter as cited above.

5 Similarly, Dacey [7] accounts for the absorption of a voter's view of the world as an inductive process.

6 On strategy-proofness see Gibbard [8] and Satterthwaite [39].

7 The strategy is printed in the operator's manual supplied with the auto. Usually, the strategy is: (1) if the oil pressure light is on (or if the oil pressure guage fails below some particular level), then stop the car; (2) if the light is off (or if the guage reads above the stated level), then continue driving. This is a trivially simple strategy. Conceptually, however, it is identical to the rather complex strategies printed in the operator's manual of a commercial jet aircraft.

8 Clearly, past events can also be considered. See Kitts [19].

9 The revised (or 'posterior') probability $P(z/y)$ is obtained via Bayes' theorem.

$$P(z/y) = \frac{P(z)\,P(y/z)}{\sum_Z P(z)P(y/z)}.$$

There are alternative probability kinematics, e.g., Jeffrey [12] and May and Harper [29].

[10] A matrix is Markov if and only if (1) each element in the matrix is non-negative, and (2) the sum of elements in each row equals unity. Note that the specification of η' presumes $P(y'/z \cap y) = P(y'/y)$, i.e., that $P(y'/y)$ is independent of z. Thus, y' is determined solely by y, but possibly with the intervention of 'noise'. On this point see [28, pp. 64–66]. There are alternative definitions of 'garbling'. See [26, p. 54, f.n. 18].
[11] Note that we are presuming that decision making is costless. This presumption is maintanined throughout the present paper.
[12] See [41] for an interesting discussion of the major contributions to decision theory in terms of the decomposition of β into ρ and μ.
[13] See [28, pp. 87–90] for an example wherein MV and NV provide reversed orderings of two information systems.
[14] See Hilpinen [11] for a review. Levi [23] presents a decision-theoretic resolution of the selection problem. Dacey [6] presents a conclusion-theoretic resolution.
[15] Balch [1] discusses the various problems mentioned here.
[16] This view is stated by Hempel as follows: "The pursuit of knowledge as exemplified by pure scientific inquiry, by 'basic research' not directly aimed at any practicial applications with corresponding utilities, is often said to be concerned with the discovery of truth" [10, p. 153]. See [10, pp. 149–163] for Hempel's elaboration of this point.
[17] This is shorthand notation for the following: $\rho(a_1, z_1) = 20$, $\rho(a_2, z_1) = 10$, $\rho(a_1, z_2) = 10$ and $\rho(a_2, z_2) = 30$.

REFERENCES

[1] Balch, M., 'On recent developments in subjective expected utility', pp. 45–55 in [3].

[2] Balch, M. and Fishburn, P., 'Subjective expected utility for conditional primitives', pp. 57–69 in [3].

[3] Balch, M., McFadden, D. L. and Wu, S. Y. (eds.), *Essays on Economic Behavior Under Uncertainty*, North-Holland Publishing Co., Amsterdam, 1974.

[4] Blackwell, D., 'Equivalent comparisons of experiments', *Annals of Mathematical Statistics* 24 (1953), 265–272.

[5] Dacey, R., 'Theory absorption and the testability of economic theory', *Zeitschrift für Nationalökonomie,* **36**: 3–4 (December, 1976), 200–215.

[6] Dacey, R., 'A theory of conclusions', *Philosophy of Science,* **45**: 4 (December, 1978), 563–574.

[7] Dacey, R., 'The role of ambiguity in manipulating voter behavior', *Theory and Decision,* **10**: 1–4 (January, 1979), 265–279.

[8] Gibbard, A., 'Manipulation of voting schemes: a general result', *Econometrica* **41** (1973), 587–601.

[9] Grunberg, E. and Modigliani, F., 'The predictability of social events', *Journal of Political Economy,* **62**: 6 (1954).

[10] Hempel, C., 'Deductive-nomological versus statistical explanation', in *Minnesota Studies in the Philosophy of Science*, Vol. 3, (H. Feigl and G. Maxwell, eds.), University of Minnesota Press, Minneapolis, 1962, pp. 98–169.

[11] Hilpinen, R., *Rules of Acceptance and Inductive Logic* (*Acta Philosphica Fennica*, Vol. 21), North-Holland Publishing Co., Amsterdam, 1968.

[12] Jeffrey, R. C., *The Logic of Decision*, McGraw-Hill, Publishing Co., Inc., New York, 1965.

[13] Hintikka, J., 'Towards a theory of inductive generalization', in Y. Bar-Hillel (ed.), *Proceedings of the 1964 International Congress for Logic, Methodology, and Philosophy of Science*, Amsterdam, 1965, pp. 274–288.

[14] Hintikka, J., 'On a combined system of inductive logic', *Studia Logico-Mathematica et Philosophica in Honorem Rolf Nevanlinna* (Fasc. 18 in *Acta Philosophica Fennica*), Amsterdam, 1965, pp. 21–30.

[15] Hintikka, J., 'A Two-Dimensional Continuum of Inductive Methods', in Hintikka and Suppes (eds.), *Aspects of Inductive Logic*, Amsterdam, 1966, pp. 113–132.

[16] Hintikka, J. and Pietarinen, J., 'Semantic information and inductive logic', in Hintikka and Suppes (eds.), *Aspects of Inductive Logic*, Amsterdam, 1966, pp. 96–112.

[17] Jeffrey, R. C., 'New foundations for a Bayesian decision theory', in Y. Bar-Hillel (ed.), *Logic Methodology and Philosophy of Science*, North-Holland Publishing Co., Amsterdam, 1965, pp. 289–300.

[18] Jeffrey, R. C., 'Frameworks for preference', pp. 74–79 in [3].

[19] Kitts, D., *The Structure of Geology*, Southern Methodist University Press, Dallas, Texas, 1977.

[20] Kyburg, Henry, 'The rule of detachment in inductive logic', in I. Lakatos (ed.), *The Problem of Inductive Logic*, North-Holland Publishing Co., Amsterdam, 1968, pp. 98–119, 'Reply', pp. 161–165.

[21] Lancaster, K., 'A new approval to consumer theory', *Journal of Political Economy*, 74 (1966), 132–157.

[22] Lancaster, K., *Consumer Demand: A New Approach*, Columbia University Press, New York, 1971.

[23] Levi, I., *Gambling with Truth*, The MIT Press, Cambridge, Mass., 1967.

[24] Luce, R. D. and Krantz, D. H., 'Conditional expected utility', *Econometrica* 39 (1971), 253–271.

[25] Marget, A. W., 'Morgenstern on the methodology of economic forecasting', *Journal of Political Economy*, 37: 3 (June, 1929), 312–339.

[26] Marschak, J., 'Economics of information systems', in M. Intriligator (ed.), *Frontiers of Quantitative Economics*, North-Holland Publishing Co., Amsterdam, 1971, pp. 32–107.

[27] Marschak, J. 'Information, decision, and the scientist', in C. Cherry (ed.), *Pragmatic Aspects of Human Communication*, D. Reidel Publishing Co., Dordrecht, Holland, 1974.

[28] Marschak, J. and Radner, R., *Economic Theory of Teams*, Yale University Press, New Haven, Conn., 1972.

[29] May, S. and Harper, W. L., 'Toward an optimization procedure for applying minimum change principles in probability kinematics', in W. L. Harper and C. A. Hooker (eds.), *Foundations of Probability Theory, Statistical Inference, and Statistical Theories of Science*, Vol. 1, D. Reidel Publishing Co., Dordrecht, Holland, 1976, pp. 137–166.

[30] Morgenstern, O., *Wirtschaftsprognose, eine Untersuchung ihrer Voraussetzungen und Moglichkeiten*, Julius Springer Verlag, Vienna, 1928.

[31] Morgenstern O., *Die Grenzen Wirtschaftspolitik*, Julius Spring Verlag, Vienna,

1934, translated by Vera Smith, as *The Limits of Economics*, W. Hodge and Co., Ltd., London, 1937. (All references will be to the latter book.)

[32] Morgenstern, O., 'Vollkommene voraussicht und wirtschaftliches gleichgewicht', *Zeitschrift für Nationalökonomie,* **6**: 3 (August, 1935), pp. 337–357, translated by Frank H. Knight as 'Perfect foresight and economic equilibrium', pp. 169–183 in [35]. (All references will be to the latter paper.)

[33] Morgenstern, O., 'Descriptive, predictive and normative theory', *Kyklos* **25** (1972), 699–714.

[34] Morgenstern, O., 'Thirteen critical points in contemporary economic thoery: an interpretation', *The Journal of Economic Literature*, **10**: 4 (December, 1972), 1163–1189.

[35] Morgenstern, O. and Schwödiauer, G., 'Competition and collusion in bilateral Markets', *Zeitschrift für Nationalökomie* **36** (1976), 217–245.

[36] v. Neumann, J. and Morgenstern, O., *Theory of Games and Economic Behavior*, Princeton University Press, Princeton, 1944.

[37] Niiniluoto, I. and Tuomela, R., *Theoretical Concepts and Hypotetico-Inductive Inference*, Reidel Publishing Co., Dordrecht, Holland, 1973.

[38] Salmon, W. C., 'Who needs inductive acceptance rules?', in I. Lakatos (ed.), *The Problem of Inductive Logic*, North-Holland Publishing Co., Amsterdam, 1968, pp. 139–144.

[39] Satterthwaite, M. A., 'Strategy-proofness and Arrow's conditions', *Journal of Economic Theory* **10** (1975), 187–217.

[40] Savage, J. L., *The Foundations of Statistics*, J. Wiley and Sons, Inc. New York, 1954.

[41] Schneeweiss, H., 'Probability and utility – dual concepts in decision theory', pp. 113–144 in G. Menges, (ed.) *Information, Inference and Decision*, D. Reidel Publishing Co., Dordrecht, Holland, 1974.

[42] Schotter, A., 'Core allocations and competitive equilibrium – a survey', *Zeitschift für Nationalökonomie* **33** (1973), 281–313.

[43] Schotter, A. (ed.), *Selected Economic Writings of Oskar Morgenstern,* New York University Press, New York, 1976.

[44] Shapley, L. S., 'The Solutions of a Symmetric Game', *Annals of Mathematics Studies* **40** (1959), 145–162.

[45] Shapley, L. S. and Shubik, M., 'Competitive outcomes in the cores of market games', *International Journal of Game Theory* **4** (1975), 229–237.

[46] Shubik, M., 'Edgeworth Market Games', *Annals of Mathematics Studies,* **40** (1959), 267–278.

LOUIS L. WILDE

ON THE USE OF LABORATORY EXPERIMENTS IN ECONOMICS

1. INTRODUCTION

Twenty years ago the focus of welfare economics began to shift from comparing allocations in a given economic environment to comparing systems of economic organization which might operate within entire classes of economic environments. This "(new)^2 welfare economics," to use Reiter's (1977) term, is not simply concerned with the outcome of a given system, but rather is concerned also with practical features of systems such as administrative feasibility, computational complexity, and cost of operation. Indeed the entire 'Public Choice' approach to welfare economics is a manifestation of this shift in emphasis. Reiter discusses a number of formal examples of this kind of work, including the well-known formulation by Ted Groves and John Ledyard of a government allocation-taxation system for the optimal provision of public goods (Groves and Ledyard, 1977).

In spite of its focus on 'real-world' considerations, most of the (new)^2 welfare economics has been theoretical. There are several reasons for this. First, the initial conceptualization of the problem was due to theorists. Their work has been very abstract and demands of the reader a substantial degree of mathematical sophistication. It has, therefore, not been generally accessible to either applied economists or decisionmakers who might be in a position to make use of the theoretical results. Second, empirical work has been difficult in this area. Often the institutions of interest do not exist and, in cases where they do, data which might allow theoretical results to be tested is hard to obtain. Finally, many of the theoretical results do not make predictions but rather suggest how institutions with desirable features ought to be designed. The needed nontheoretical work is therefore more along the lines of implementation than testing.

One possible solution to the absence of nontheoretical work in the (new)^2 welfare economics is nonlaboratory experiments. However, this solution has difficulties of its own. Nonlaboratory experiments tend to be very expensive, often prohibitively so. They also require the cooperation of individuals whose major priority is making decisions, not contributing to economics at a basic scientific level.[1]

J. C. Pitt (ed.), Philosophy in Economics, 137–148.

Another possible solution to the problem, and the one which is of primary interest here, is laboratory experiments. The use of laboratory experimental techniques in economics has increased dramatically in recent years, and virtually all of the major economics journals have begun to publish the results of laboratory experiments. This increased interest has been a direct response to the absence of nontheoretical work in the (new)^2 welfare economics. The purpose of this paper is to review, and develop when needed, the foundations of the use of laboratory experimental techniques in economics (as propounded by economists). This task will be undertaken in the next section of the paper. The discussion there will be based on the work of Charles Plott and Vernon Smith, although some modifications of their 'percepts' and 'axioms' will be necessary. A third section will briefly discuss limitations of laboratory experiments.[2]

2. A PRECEPTUAL FOUNDATION

In the introduction it was suggested that the use of laboratory experimental techniques in economics has arisen in response to the need for a nontheoretical (new)^2 welfare economics. One reason for taking this point of view is that even though economists are interested in both individuals and institutions, systematic rationalizations of the use of laboratory experiments by economists have been developed only for the latter. Furthermore, a number of recent laboratory experiments have been designed to study the implementation of particular theoretical systems. For example, Smith (1976) and Ferejohn, Forsythe and Noll (1979) both analyze practical aspects of decentralized decisionmaking systems for public goods such as those formulated by Groves and Ledyard (1977). In spite of this work, thought on the subject is still in its infancy and is being continually refined as practical experience accumulates. In fact, the 'literature' consists primarily of two papers, one by Vernon Smith (1977) and the other by Charles Plott (1978). This paper will draw heavily on the work of Plott and Smith. However, several significant modifications of the foundations offered by those authors will be introduced.

The basic idea behind a laboratory experiment in economics is to create a small-scale microeconomic environment in the laboratory where adequate control can be maintained and accurate measurement of relevant variables guaranteed. The definition of a microeconomic environment which I will use starts with Reiter's description of an "economic environment." An economic environment consists of a list of agents $\{1, \ldots, n\}$ and commodities $\{1, \ldots, \ell\}$. Each agent is described by a preference relation α_i (represented generally by a utility function u^i), a technology T^i (represented generally by a production possibilities set) and an initial endowment vector, ω^i. The ith agent is thus described by a triple, $e^i = (\alpha_i, T^i, \omega^i)$. Commodity space is taken to be $\mathbb{R}^\ell$. Given that the list of commodities is fixed, the economic environment is then described by $e = (e^1, \ldots, e^n)$.

To complete a microeconomic environment one needs to specify the institutional setting. As a theoretical matter, this is the focus of the exercise. For our purposes, a very simplified formalization will be adequate. Let individual i have the opportunity to make a decision $d_i \in D_i$. Each individual is assumed to select d_i so as to maximize u^i (note that externalities are not rules out). An institution is defined by the collection of sets $D = D_1 \times \ldots \times D_n$ and a mapping $I : D \rightarrow \mathbb{R}^{\ell n}$ which takes decisions into final allocations. A microeconomic environment is then described by $E = (e, I)$.

The essential features of the above construction are as follows. First, a microeconomic environment is composed primarily of two elements, a collection of individuals and an institutional setting. Second, two properties characterize the individuals; they are assumed to possess consistent preferences and to make decisions so as to maximize their own well-being. Third, these decisions act through the institutional setting in order to determine final outcomes.

A laboratory experiment in economics attempts to create and study a small-scale microeconomic environment. Its purpose is to uncover systematic relationships between individual preferences, institutional parameters, and outcomes.[3] To realize this purpose, the experimenter must have control over both the preferences of the individuals participating in the experiment and the institutional parameters which govern final allocations. Thus, the initial tasks of the experimenter are to ensure that a genuine microeconomic environment has been created, and to ensure that enough control over individual preferences and institutional parameters can be maintained to uncover any systematic relationships between them and final outcomes. It is my aim here to find a set of sufficient conditions which, if satisfied, will guarantee that these requirements have been met.

Consider first the control of individual preferences. Sufficient conditions for such control are based on the theory of 'induced value' as developed by Smith (1976, 1977). Smith's theory is implemented by mapping final allocations into a reward structure. If the reward structure satisfies certain properties (detailed below) then adequate control over preferences can be guaranteed.

Smith identifies three 'precepts' as constituting a foundation for the use of laboratory experiments in the study of resource allocation mechanisms. These are *nonsatiation, complexity*, and *parallelism*. The term precept is adopted here directly from Smith. Apparently he uses it to suggest that his three conditions are properties only intended to connote general rules of action related to proper experimental design. They are thus not to be regarded as self-evident truths (as the term axiom might imply) or as strictly sufficient conditions for a valid experiment. Indeed, Smith never states in any general way the objectives which govern the design of laboratory experiments, so that it is impossible to interpret the precepts as sufficient conditions for a 'valid' experiment. Nevertheless, they provide an extremely useful starting point for developing such a set of sufficient conditions.

According to Smith, nonsatiation requires that

> given a costless choice between two alternatives, identical except that the first yields more of the reward medium (e.g., currency) than the second, the first will always be chosen (preferred) over the second, by an *autonomous* individual.[4]

Complexity recognizes that

> in general individual decision makers must be assumed to have multidimensional values which attach nonmonetary subjective cost or value to (1) the process of making and executing individual or group decisions, (2) the end result of such decisions, and (3) the rewards (and perhaps behavior) of other individuals involved in the decision process.[5]

Parallelism asserts that

> propositions about the qualitative behavior of individuals and of markets and other resource allocation mechanisms that have been tested in laboratory experiments apply also to nonlaboratory environments where similar *ceteris paribus* conditions prevail.[6]

Smith's precepts are intended to guarantee that a well-defined microeconomic environment has been created in the laboratory, that adequate control over that environment can be maintained, and that any 'results' obtained are relevant outside the laboratory. Consider the first of these goals. A well-defined microeconomic environment requires that individuals have consistent preferences and act so as to maximize their own well-being. Nonsatiation guarantees that both these requirements will be satisfied. However, a well-defined microeconomic environment also requires that individual decisions act through an institutional setting in order to determine final allocations which, in turn, determine rewards. In other words, rewards earned by individuals must be tied to their decisions. None of Smith's precepts guarantee that this requirement is satisfied. Thus a fourth precept, which I will call *saliency*, must be added to Smith's precepts. Saliency requires that the reward earned by an individual is tied to decisions made by that individual.

Together saliency and nonsatiation guarantee that the requirements of a well-defined microeconomic environment are satisfied. Saliency implies that the amount of the reward medium earned is linked to the decisions made by the subjects and nonsatiation implies that the amount of the reward medium earned is always important to the subjects. Thus a reward structure which satisfies saliency and nonsatiation can also be used to control preferences systematically. An example of a particular experiment will make this clear.

Consider an experimenter who wishes to study repeated oral double auction markets. An oral double auction is a very simple institution. A single homogeneous good is to be bought and sold. There are two types of agents, buyers and sellers. All buyers can make bids and all sellers can make offers. These bids and offers are displayed for all agents to see. If a bid is made by a buyer and is acceptable to a seller, that seller simply accepts the bid and a transaction is recorded. Similarly, if an offer is made by a seller and is acceptable to a buyer, that buyer simply accepts the offer and a transaction is recorded. All agents observe the prices at which transactions are made.

In order to study this institution experimentally, it is necessary to 'induce' preferences which will generate desired market demand and supply curves. On the demand side, subjects are given a table listing monetary rewards to be provided by the experimenter for units of the commodity bought. The subjects then earn the difference between the total 'redemption value' of the units they have bought and the costs incurred in their purchase. As long as the subjects prefer more money to less, the redemption values specified by the experimenter constitute a well-defined demand curve. Similarly, on the supply side, subjects are given a table listing monetary costs to be assessed by the experimenter for units of the commodity sold. The subjects then earn the difference between the total revenue they have collected from sales and the total cost of the units they have sold. Again, as long as the subjects prefer more money to less, the costs specified by the experimenter constitute a well-defined supply curve. Shifts in the demand curve can be accomplished by changing the redemption values. Shifts in the supply curve can be accomplished by changing the initial costs. Moreover, the same references can be induced on different subject pools by using identical redemption values or initial costs.

Consider nonsatiation and saliency in terms of this example. Nonsatiation is satisfied by the reward structure because it is reasonable to assume that subjects prefer more money to less. Saliency is satisfied by the reward structure because buyers earn the difference between the redemption value and the sale price while sellers earn the difference between the sale price and the initial cost. Hence buyers have an incentive to minimize the sale price and sellers have an incentive to maximize it.

Using this process of inducing values to control preferences is not without its difficulties. Problems can arise from two distinct sources. First, subjects may place subjective valuations on their participation in the experiment over and above any direct payoff they receive in terms of the reward medium. Some subjects may enjoy making the calculations and decisions required of them by the experiment, while others may find such activity arduous. Second, subjects may also place subjective valuations on the rewards earned by other participants in the experiment. For example, equity may be of great concern to some subjects. The former effect is usually controlled by using a reward structure with high payoffs. I will refer to this as *dominance* of the reward structure. The latter effect is usually controlled (when appropriate) by keeping subjects uninformed of the overall pattern of the reward structure across individuals. I will refer to this as *privacy* of the reward structure. If the reward structure satisfies dominance and privacy then Smith's precept of complexity, which is stated more as a cautionary warning than as a rule of action, will have been rendered moot.

Consider again the oral double auction market described above. Dominance is satisfied by the reward structure because a 'commission' is usually paid on all transactions (ensuring that marginal units will be traded) and privacy is satisfied by the reward structure because against see only bids and offers, not redemption values and initial costs. (Often they are isolated at computer terminals and thus never come into contact with each other).

Suppose now that the reward structure satisfies saliency, nonsatiation, dominance, and privacy. Then, in fact, the laboratory experiment constitutes a small-scale microeconomic environment in which real economic agents make real economic decisions. Moreover, these decisions are based on values controlled by the experimenter. Control is crucial because it is necessary for measurement and thus replicatability. Replicatability, in turn, allows the experimenter to identify *systematic* relationships between preferences, institutional parameters and outcomes. These systematic relationships constitute the 'results' of laboratory experiments in economics. In other words, any systematic relationship between preferences, institutional parameters, and outcomes which has been identified by replication is *by definition* a result. The final question is, of what use are these results?

To date three distinct uses of laboratory experimental techniques in economics have been discussed. They are: (a) the appropriateness of competing theories can be distinguished; (b) theories which are clearly nonsense can be exposed; and (c) experience can be obtained with new modes of organization. These three cases have been analyzed in depth by Plott (1978) where several examples of each are provided. Herein only a cursory discussion will be offered.

The use of laboratory experiments to distinguish between competing theories is perhaps the most classic. As Smith puts it, "the best experiment is the *crucial* experiment whose outcome clearly distinguishes between competing theories."[7] The problem, of course, is that the conditions which define the crucial experiment rarely occur naturally. But they can often be created in the laboratory.

It is important to note here that a crucial experiment need not be 'realistic' in the sense that nonlaboratory experiments are. It only needs to include those parameters relevant to the theories which are to be tested. If it does (assuming that saliency, nonsatiation, dominance, and privacy are satisfied by the reward structure) then any failure to distinguish between competing theories is the fault of the theories, not the experiment.

This point is valid with respect to the use of laboratory experiments to reject some theories as nonsense, too. That is, if an experiment includes all parameters relevant to a particular theory, and if the theory fails to predict well in the simplified setting of the laboratory, then it cannot be expected to predict well in more complex environments. Again, the only requirements needed to reach this conclusion are that saliency, nonsatiation, dominance, and privacy are satisfied by the reward structure. The experiment does not need to be 'realistic' and no presumptions need be made about its connection to more complex ('real-world') environments.

The third use of laboratory experiments, to gain experience with new modes of organization, is different in this regard. Instead of starting with an existing mode of organization and trying to develop and test theories related to it, the researcher starts with a theoretical concept which is "devoid of operational detail."[8] The task is to construct a new mode of organization which exhibits the desired performance features. Plott has referred to this as "institutional engineering."[9] It is precisely the nontheoretical (new)^2 welfare economics discussed in the introduction of this paper. For this use of laboratory experiments to be valid, more than saliency, nonsatiation, dominance, and privacy of the reward structure is required. There must be a link to nonlaboratory environments. This is where Smith's precept of parallelism becomes important.

Recall that parallelism asserts that

> propositions about the qualitative behavior of individuals and of markets and other resource allocation mechanisms that have been tested in laboratory experiments apply also to nonlaboratory environments where similar *ceteris paribus* conditions prevail.[10]

In terms of the definitions introduced above, this precept might be rephrased as asserting that the results of laboratory experiments apply to any microeconomic environment where similar *ceteris paribus* conditions prevail. The reasons for this rephrasing are twofold. First, "propositions about the qualitative behavior of individuals and of markets and other resource allocation mechanisms that have been tested in laboratory experiments" are generally statements concerning relationships between preferences, institutional paramaters, and outcomes. The only such statements which are valid are those which can be replicated. Such statements, however, constitute the 'results' of any experiment. Second, the distinction between laboratory and nonlaboratory environments is irrelevant. What is important is that the results apply to *any* microeconomic environment where similar *ceteris paribus* conditions prevail. Here it is important to recognize that "similar *ceteris paribus* conditions" prevailing does not mean identical environments. It means in part that parameters affecting preferences or institutional structure are held constant except those relating to the behavioral propositions under study. These constant parameters need not be identical.

Parallelism is crucial to justifying the third and potentially most fruitful use of laboratory experimental techniques in economics listed above. As such, it is by no means noncontroversial. In fact, a major criticism of laboratory experiments is that they *are* unrealistic. That is, individuals and institutions in the real-world are claimed to be much more complex than their laboratory counterparts, and therefore any results obtained through laboratory experiments have no relevance to real-world behavior. Such criticisms either attack the theory of induced value as embodied in the precepts of saliency, nonsatiation, dominance and privacy or they attack the precept of parallelism. Smith offers two responses to the latter.

> First, if the purpose of the experiment is to test a theory, are the elements of alleged unrealism in the experiment parameters of the theory? If not then the criticism must be directed to the theory as much as to the experiment. Laboratory experiments are normally as 'rich' as the theories they test. Secondly, are there field data to support the criticism, i.e., data suggesting that there may be differences between laboratory and field behavior. If not, then the criticism is pure speculation; if yes, then it is important to parameterize the theory to include the behavior in question.[11]

Attacks on the theory of induced value are primarily questions of methodology since the theory itself is internally consistent.[12] But in this case replication is the key. If replicable relationships between preferences, institutional parameters, and outcomes have been identified, then the researcher must extend or modify existing theory to explain the relationships. After all, it is only necessary that saliency and nonsatiation be satisfied in order that a microeconomic environment exist. Everything else is simply a matter of control and measurement.

The modifications of Smith's preceptual foundation introduced above were necessary to ensure that a genuine microeconomic environment was created and to ensure that enough control over individual preferences and institutional parameters could be maintained to uncover any systematic relationships between them and outcomes. One final problem with Smith's preceptual foundation is that the concept of "similar *ceteris paribus* conditions" is not defined in a precise way. It is here that Plott (1978) becomes particularly relevant.

As in this paper, Plott starts with Smith's theory of induced value, introducing two additional 'axioms'.[13] Axiom one is that "the relationship between outcomes, preferences and institutions are (supposed to be) independent of the social alternatives."[15]

Axiom one essentially asserts that relationships between preferences, institutional parameters, and outcomes are independent of the sources of preferences. To paraphrase Plott's example, it is irrelevant whether a person is willing to pay ten dollars maximum for a shirt because (a) he or she thinks it is pretty, (b) a friend thinks it is pretty, or (c) it tastes good. In each case, that person's contribution to the total market demand for shirts is the same and, hence, so is the equilibrium price of shirts.

Axiom two essentially asserts that 'labels' don't matter. Thus a competitive market for wheat should exhibit the same qualitative properties as a competitive market for health services.

These two axioms are closely related to parallelism. In particular, they are necessary for parallelism but not sufficient. In essence, they begin to define "similar *ceteris paribus* conditions" in a precise way. Ultimately, we would like to find a set of necessary and sufficient conditions which fully characterize the concept of *ceteris paribus* in the context of parallelism. Whether such a set of 'axioms' can be found is an unresolved issue. Nevertheless the notion of *ceteris paribus* is crucial to all of economics, and thus efforts to find such a set of 'axioms' is of benefit to the entire profession, not just experimentalists.[16]

3. LIMITATIONS OF THE USE OF LABORATORY EXPERIMENTAL TECHNIQUES IN ECONOMICS

So far in this paper, a set of sufficient conditions has been identified which, if satisfied, guarantee that laboratory experimental techniques are appropriate nontheoretical tools for economists. Nonsatiation and saliency of a reward structure ensure that a well-defined microeconomic environment has been created, dominance and privacy of a reward structure ensure that enough control over individual preferences and institutional parameters can be maintained to uncover any systematic relationships between them and outcomes, and parallelism ensures that any results obtained are useful outside the laboratory. Nothing yet has been said about when these conditions are likely to be satisfied.

It is not my intention to draw specific conclusions regarding this matter, but rather to make the point that criticisms of laboratory experiments must focus on precisely this issue. Section 2 of this paper indicates that laboratory experimental techniques can be valid tools for economists. The relevant questions are whether the sufficient conditions identified in this paper are ever likely to be satisfied in general or, more importantly, whether they are in fact satisfied in particular cases. An argument that they are not likely to be satisfied in general seems implausible, but there are undoubtedly examples of particular cases in which they fail.

Economists have gained experience with a wide spectrum of experiments in recent years, from relatively simple bargaining experiments (Roth and Malouf, 1979), to the oral double auctions described in the last section,[17] to relatively complicated market experiments such as Hong and Plott (1977) (which analyzed the implications of rate filing for the domestic dry bulk transportation industry on inland U.S. waterways) and Plott and Wilde (1979) (which analyzed markets for professional diagnosis and services). These experiments are listed seemingly in order of those most likely to satisfy the sufficient conditions of Section 2 to those least likely to do so, given the present level of understanding. This assertion is premature, however, because laboratory experimental techniques are new to economists. They do hold much promise, though, and such strong judgments should really be suspended until more research has been done.

California Institute of Technology

NOTES

* I would like to thank Ed Green, Charles Plott, Alan Schwartz, and Vernon Smith for helpful comments on the first draft of this paper.

[1] Nonlaboratory experiments include token economies as well as 'real-world' economies. For a review of the former see Tarr (1976), for an example of the latter see Manning, Mitchell and Action (1976).

[2] Animals have been used as subjects in some laboratory experiments by economists. Issues related to the use of such experiments in economics go well beyond this paper. See Kagel *et al.* (1975) for examples of this type of research.
[3] This statement is adopted from Plott's 'fundamental equation' (Plott, 1978). It is nevertheless controversial. Uses of laboratory experimental techniques will be discussed in more detail later in this paper.
[4] Vernon Smith (1977), p. 3.
[5] *Ibid.*, p. 5.
[6] *Ibid.*, p. 7.
[7] *Ibid.*, p. 8.
[8] Charles Plott (1978), p. 36.
[9] The choice of this label was certainly not accidental. Notice the relationship between the use of laboratory experiments and 'demonstration' projects in large-scale R & D.
[10] Smith (1977), p. 7.
[11] *Ibid.*, p. 8.
[12] One point of this paper is that the 'theory' of induced preferences is not really a theory at all.
[13] The term 'axiom' is Plott's. He apparently uses it because these properties are true of all economic models.
[14] Plott (1978), p. 3.
[15] *Ibid.*, p. 6.
[16] In commenting on an earlier draft of this paper, Vernon Smith also pointed out the need to articulate what *ceteris paribus* might mean in the context of parallelism. This is a complicated and important issue, but not one I intend to resolve in this paper.
[17] This is perhaps the best understood experimental institution. A number of variants of the basic oral double auction have also been studied (e.g., Isaac and Plott, 1978) and some sophisticated theoretical work has also been done *in response* to this work (Easley and Ledyard, 1979).

REFERENCES

Easley, D. and Ledyard, J., 'A theory of double auctions', presented at NSF Conference on Experimental Economics, Tucson, Arizona (October 1979).

Ferejohn, J., Forsythe, R., and Noll, R., 'Practical aspects of the construction of decentralized decision making systems for public goods', in C. Russel (ed.), *Applications of Social Choice Theory*, Resources for the Future, Washington, D.C., 1979.

Groves, T. and Ledyard, J., 'Optimal allocation of public goods: a solution to the free-rider problem', *Econometrica* 45 (May 1977), 783–810.

Hong, J. and Plott, C., 'Implications of rate filing in domestic dry bulk transportation on inland waters: an experimental approach', Caltech Social Science Working Paper No. 164, July 1977.

Isaac, M. and Plott, C., 'Price controls and the behavior of auction markets: an experimental examination', Caltech Social Science Working Paper No. 253, February 1979.

Kagel, J. H., Battalio, R. C., Rachlin, H., Green, L., Basmann, R. L., and Klemin, W. R., 'Experimental studies of consumer demand behavior using laboratory animals', *Economic Inquiry* (March 1975), 22–38.

Manning, W. G., Mitchell, B. M., and Action, J. P., 'Design of the Los Angeles peak load experiment for electricity', Report No. R-1955-DWP (November 1976), The Rand Corporation, Santa Monica, Ca.

Plott, C., 'The application of laboratory experimental methods to public policy', Caltech Social Science Working Paper No. 223 (July 1978). Forthcoming in a *Resources for the Future* volume, Clifford Russel, editor.

Plott, C. and Wilde, L., 'Professional diagnosis vs. self-diagnosis: an experimental examination of some special features of markets with uncertainty', unpublished manuscript, Caltech (October 1978).

Reiter, S., 'Information and performance in the (new)^2 welfare economics', *American Economic Review* **67** (February 1977), 226–234.

Roth, Alvin and Malouf, M. K., 'Game-theoretic models and the role of information in bargaining', unpublished MS, University of Illinois (January 1979).

Smith, V., 'Experimental economics: induced value theory', *American Economic Review* **66** (May 1976), 274–279.

Smith, V., 'Relevance of laboratory experiments to testing resource allocation theory', presented at the Conference on Criteria for Evaluation of Econometric Models, June 1977. Forthcoming in *Evaluation of Econometric Models*, J. Kimenta and J. Ramsey, (eds.).

Smith, V., 'Incentive compatible experimental processes for the provision of public goods', in V. Smith (ed.), *Research in Experimental Economics*, vol. 1 (JAI Press, 1978).

Tarr, D. G., 'Experiments in token economies: a review of the evidence relating to assumptions and implications of economic theory', *Southern Economic Journal* (October 1976), 136–143.

I. J. GOOD*

SOME LOGIC AND HISTORY OF HYPOTHESIS TESTING

1. INTRODUCTION

The foundations of statistics are controversial, as foundations usually are. The main controversy is between so-called Bayesian methods, or rather neo-Bayesian, on the one hand and the non-Bayesian, or 'orthodox', or sampling-theory methods on the other.[1] The most essential distinction between these two methods is that the use of Bayesian methods is based on the assumption that you should try to make your subjective or personal probabilities more objective, whereas anti-Bayesians act as if they wished to sweep their subjective probabilities under the carpet. (See, for example, Good (1976).) Most anti-Bayesians will agree, if asked, that they use judgment when they apply statistical methods, and that these judgments must make use of intensities of conviction,[2] but that they would prefer not to introduce numerical intensities of conviction into their formal and documented reports. They regard it as politically desirable to give their reports an air of objectivity and they therefore usually suppress some of the background judgments in each of their applications of statistical methods, where these judgments would be regarded as of potential importance by the Bayesian. Nevertheless, the anti-Bayesian will often be saved by his own common sense, if he has any. To clarify what I have just asserted, I shall give some examples in the present article.

My own philosophy of probability and statistics is a Bayes/non-Bayes compromise. I prefer to call it the Doogian philosophy rather than the Good philosophy because the latter expression might appear self-righteous. Although I have expounded this philosophy on several previous occasions (for example, Good, 1950, pp. 31–32; 1952; 1962a; 1965, pp. 6–11; 1971) it can be roughly expressed succinctly enough to justify describing it yet again. In fact, the theory can be expressed so succinctly that it is liable to be ignored. The main influences on me were acknowledged in the Preface of Good (1950).

Let E, F, G, H denote propositions and let $P(E|F)$ (etc.) denote real numbers statisfying the usual probability axioms. Let $P'(E|F) \geqslant P'(G|H)$ mean that your intensity of conviction in E given F is at least as great as that in G given H. Such a comparison can be used as input into a 'black box' in which the primes are omitted with the result that the comparison is converted

J. C. Pitt (ed.), Philosophy in Economics, 149–174.

into a numerical inequality. (The numerical values of $P(E|F)$, etc., need not be known to you nor to any one else.) Sometimes you cannot compare two intensities of conviction, that is, your intensities of conviction are only partially ordered (not completely ordered). In the black box the usual axioms, in which probabilities are regarded as numbers, are used to produce new inequalities as output. Primes are then reinstated and thus 'discernments' are produced to be appended to your 'body of beliefs'. The purpose of the theory is to enlarge your body of beliefs and to check it for consistency. When you find an inconsistency in your body of beliefs you should try to remove it if the cost of doing so is not too great.

By thinking about uncontroversial numerical probabilities associated with idealized games of chance, you can introduce 'landmark' probabilities into your body of beliefs. Then each intensity of conviction can be said to belong to some interval of real numbers whose endpoints are called lower and upper probabilities. Having understood the theory to this point it becomes unnecessarily pedantic to preserve the primes and you can use the symbol $P(E|F)$ in two senses. It can also be call 'the' probability although it is not necessarily a definite real number. One can deduce from this simple theory a set of axioms for upper and lower probabilities if one wishes (Good, 1962a).

A similar theory incorporating partially ordered utilities can then be constructed in a natural, almost an obvious, way (Good, 1952). Yet the obvious is often overlooked.

A theory of partially ordered probabilities is in a sense a compromise between a 'strict' or 'sharp' Bayesian philosophy in which all probabilities are precise, and non-Bayesian philosophies in which they are assumed merely to lie between 0 and 1 (Good, 1976, p. 137).

Any sufficiently complete theory can be expressed in terms of axioms, rules and suggestions. I stated a set of axioms carefully in Good (1950, pp. 19 and 49) and will not repeat them here except to mention the product axiom. This is

$$P(E \,\&\, F|H) = P(E|H) \,.\, P(F|E \,\&\, H)$$

and its meaning is that if it is assumed that any two of the probabilities belong to certain intervals, then the third probability can be inferred to belong to some interval by using the equation. In my opinion this single assumption discriminates between Bayesians and non-Bayesians as effectively as any other equally simple assumption.

For a codification of the rules and suggestions see Good (1971). Unlike a system of axioms, such a codification cannot be complete, and it does not readily lead to mathematical theorems, but I believe it is very useful.

As a psychological aid to introspection, for eliciting your own probabilities, I advocate the use of probability distributions of probabilities. This gives rise to a hierarchical theory the history of which is reviewed in some detail in Good (1979). It shows how a good philosophical point of view leads to practical statistical procedures, a possibility that might surprise many philosophers and statisticians. Like mathematics, philosophy can be either pure, applied, or applicable.

This survey of 'Doogianism' has of course been highly succinct and the reader with an open mind who would like to see more details, if such a reader exists, will presumably examine some of the references.

This philosophy will act as the underpinning, in this paper, of the discussion of hypothesis testing. It is familiar that the probability of the outcome of an experiment is usually exceedingly small, especially when described in detail, and cannot be used by itself for hypothesis testing because it has no intuitive appeal. We shall describe three methods for producing a criterion of more sensible size, namely the use of (i) tail-area probabilities, (ii) Bayes factors, or their logarithms the weights of evidence, and (iii) surprise indexes. These concepts, and others mentioned in this introduction, will be explained later. A "paradox" pointed out by Jeffreys, Good, and Lindley constitutes a serious logical objection to the first method for large samples, if the Likelihood Principle is accepted, and there are some difficulties in applying the second and third methods. These problems perhaps force one to adopt a Bayes/non-Bayes compromise or synthesis, and this synthesis will be exemplified.

2. EVIDENCE

Does your degree of belief (= intensity of conviction = subjective probability = personal probability) in some hypothesis or theory or proposition H depend only on the evidence E? It depends what you mean by the evidence. Your belief that 43 917 = 43 917 depends on logic and conventions rather than on the result of an experiment. Similarly, your judgment of the probability of a hypothesis depends on its prior probability as well as on the results of some experimental trial. If some one guesses a single decimal digit correctly you wouldn't seriously believe that he could always do so, for you would judge the prior probability of this hypothesis as being too small.

The usual way to change initial or prior probabilities into final or posterior probabilities is by means of Bayes's theorem. For a convenient formulation of this theorem it is customary to introduce the term 'likelihood' in a sense first used by R. A. Fisher in a special case. The probability of E given H, denoted by $P(E|H)$, is also called the likelihood of H given E. The special case considered by Fisher is when the value of $P(E|H)$ is uncontroversial, in fact 'tautological'; that is, where the numerical value of $P(E|H)$ is determined by the *definition* of H (in which case H is called a *simple statistical hypothesis*). For example, H might state that the probability of r successes in some experiment is $e^{-r}a^r/r!$; then of course the observation of r successes has this probability, given H, by definition. A Bayesian is prepared to use the notation $P(E|H)$ for probabilities that are not tautological, even for subjective (personal) probabilities, so, for a Bayesian, the concept of likelihood is more widely applicable than for a non-Bayesian such as Fisher. (At least he *thought* he was a non-Bayesian.) It might be advisable to call $P(E|H)$ the 'Bayesian likelihood' of H given E when this probability is not tautological. (One reason for this caution is that the expression 'likelihood ratio' is usually used by statisticians in a technical sense as a ratio of maximum likelihoods.)

The concept of likelihood or Bayesian likelihood is used only when there are at least two hypotheses, so let's suppose we have n mutually exclusive hypotheses $H_1, H_2, \ldots, H_n$. Then we have n likelihoods $P(E|H_1), \ldots, P(E|H_n)$. In some usages any set of probabilities $kP(E|H_1), \ldots, kP(E|H_n)$ is called a set of likelihoods, where k is any positive constant. With either of these definitions we can express Bayes's theorem, in the form suggested by Jeffreys as "final probability is proportional to intial probability $\times$ (Bayesian) likelihood" or, in symbols

$$P(H_i|E) \propto P(H_i) \,.\, P(E|H_i).$$

Note the immediate deduction

$$\frac{P(H_i|E)}{P(H_j|E)} = \frac{P(H_i)}{P(H_j)} \cdot \frac{P(E|H_i)}{P(E|H_j)} \,.$$

In particular, if $\bar{H}$ denotes the negation of H, we have

$$O(H|E) = O(H) \,.\, P(E|H)/P(E|\bar{H}),$$

where $O(H|E)$ means $P(H|E)/P(\bar{H}|E)$ and is called the *odds* of H given E. The ratio $P(E|H)/P(E|\bar{H})$ is called the *(Bayes) factor in favor of H provided by E* and may be denoted by $F(H{:}E)$. (Jeffreys, 1939, denoted it by K but did not given it a name.) The logarithm of the Bayes factor was independently called the *weight of evidence* in favor of H provided by E by C. S. Peirce (1878), by Good (1950) (and in about forty other publications by him) and by Minsky and Selfridge (1961). The weight of evidence can be added to the initial log-odds of the hypothesis to obtain the final log-odds. If the base of the logarithms is 10, the unit of weight of evidence was called a *ban* by Turing (1941) who also called one tenth of a ban a *deciban* (abbreviated to db). Like the decibel in acoustics, the deciban is about the smallest weight of evidence perceptible to the human mind. I hope that one day judges, detectives, doctors and other earth-ones will routinely weigh evidence in terms of decibans because I believe the deciban is an intelligence-amplifier.

If someone guesses one digit correctly the hypothesis that he can always do so gains a Bayes factor of ten and, as I said before, you still don't believe the hypothesis. But if a new typewriter prints the digit corresponding to a key you press, you would tend to believe it will always do so until it breaks down. Therefore there is simply no question of the likelihoods alone determining your states of belief in the various possible hypotheses. They can only modify the beliefs you had before you made the relevant observation.

Nevertheless, there is something called the 'likelihood principle', and, if it is correctly stated, it is a trivial consequence of Bayes's theorem. It can be stated in the following manner: *Let E and E' be two distinct experimental results or observations. Suppose that they do not affect the utilities (if true) of hypotheses $H_1, H_2, \ldots, H_n$ under consideration. Suppose further that E and E' provide the same likelihoods to all the hypotheses, that is, that $P(E|H_i) = P(E'|H_i)$ $(i = 1, 2, \ldots, n)$. Then E and E' should affect your beliefs, recommendations, and actions concerning $H_1, H_2, \ldots, H_n$ in the same way.* Clearly this principle is built into Bayes's theorem. When $n = 2$ the principle is built into the terminology 'weight of evidence'.

The likelihood principle has been supported by some non-Bayesians who want to avoid the use of probabilities that are neither apparently physical (material) nor tautological (mathematical) but are subjective (personal) or are credibilities (logical probabilities). Unfortunately members of the likelihood brotherhood have sometimes given the impression that the likelihoods by themselves always enable you to choose between hypotheses.[3] We have already seen, by the example of the guessed digit, that the likelihoods by themselves are clearly not enough for this purpose (although they are 'sufficient statistics' in Fisher's technical sense). They tell you all you need to know about the experimental results, if your model of the situation is not to be questioned, but the result of an experiment is not by itself enough for choosing between hypotheses. If it were, there would be no need for the Duns–Ockham razor. As de Finetti (1975, p. 248) said,

> . . . they ignore one of the factors (the prior probability) altogether, and treat the other (the likelihood) as though it . . . meant something other than it actually does. This is the same mistake as is made by someone who has scruples about measuring the arms of a balance (having only a tape-measure at his disposal . . .), but is willing to assert that the heavier load will always tilt the balance (thereby implicitly assuming, although without admitting it, that the arms are of equal length!).

We never reject a hypothesis H merely because an event E of very small probability (given H) has occurred although we often carelessly talk as if that were our reason for rejection. If the result E of an experiment or observation is described in sufficient detail its probability given H is nearly always less than say one in a million. As Jeffreys (1939, p. 315) said

> If mere improbability of the observations, given the hypothesis, was the criterion, any hypothesis whatever would be rejected. Everybody rejects the conclusion [of rejecting hypotheses because of improbability of the evidence] but this can only mean that the improbability of the observations, given the hypothesis, is not the criterion, and some other must be provided.

If we want to be able to say that H should be rejected "because the observation is too improbable" given H we have to do more than compute $P(E|H)$ even when this probability can be computed. Let us consider various approaches to this problem.

3. THE TESTING OF HYPOTHESES

The Bayesian approach. A Bayesian believes it is meaningful to talk about the (Bayes) factor against (or in favor of) H, or about its logarithm the weight of evidence. In statistical applications H is often (at least as an approximation) a simple statistical hypothesis in the sense that $P(E|H)$ is a tautological probability while $\bar{H}$ is often composite, that is, it is a logical disjunction of a set of simple statistical hypotheses. For the sake of simplicity of exposition I shall suppose that $\bar{H}$ is the logical disjunction "H_1 or H_2 or H_3 or . . . " (although the number of such hypotheses is often non-countable in statistical models), where $H_1, H_2, H_3, \ldots$ are mutually exclusive. (In other words $\bar{H}$ is true if and only if one of $H_1, H_2, H_3, \ldots$ is true.) The probabilities $P(H_i|\bar{H})$ and $P(E|H_i)$ are assumed to exist and $P(E|\bar{H})$ can then 'in principle' be calculated by the formula

$$P(E|\bar{H}) = P(E|H_1)P(H_1|\bar{H}) + P(E|H_2)P(H_2|\bar{H}) + \ldots.$$

Then the Bayes factor $P(E|\bar{H})/P(E|H)$, against the null hypothesis, can in principle be calculated, and its logarithm is the weight of evidence against the 'null hypothesis' H provided by E, say $W(\bar{H}:E)$. Note that this is mathematically independent of the initial or prior probability $P(H)$. The main objection to this Bayesian approach is of course that it is usually difficult to specify the probabilities $P(H_i|\bar{H})$ with much precision. The Doogian reply is that we cannot dispense with judgments of probabilities and of probability inequalities. Also that non-Bayesian methods also need Bayesian judgments and merely sweep them under the carpet.

The tail-area-probability approach. In order to judge H in terms of a probability that is not necessarily microscopic, 'tail-area probabilities' (sometimes called P-values) are often used. A tail-area probability is, by definition, the probability that the outcome of an experiment or observation would have been 'at least as extreme' as the actual outcome E.

Different statisticians have interpreted a tail area probability P in different ways. Some statisticians say that if P is small enough, then the null hypothesis H should be rejected. Braithwaite (1951) argued that rejection of a hypothesis is always provisional and that there are 'degrees of provisionality' of rejection. Many statisticians decide in advance of an experiment what threshold P_0 should be used such that if $P < P_0$ then H should be 'rejected', and they don't mention whether the rejection is only provisional. Typical values of P_0 are 0.05, 0.02, 0.01, 0.005, 0.002, and 0.001 because Fisher had issued tables (for some random variables) corresponding to these thresholds. I understand that he issued the tables in this form partly to avoid having to use Karl Pearson's tables. The threshold would be chosen depending partly on the initial probability that H is true, or is an adequate approximation, and partly on the seriousness of rejecting H when it is true or accepting it when false. Some statisticians choose a threshold without being fully conscious of why they chose it. Others will put one asterisk against outcomes that have $0.05 > P > 0.01$, two asterisks when $0.01 > P > 0.001$, and three asterisks when $P < 0.001$. There is no special justificaton for doing this except that it is conventional and saves work if tables like Fisher's are being used. If P can be calculated at little cost, which is increasingly possible, it seems better to state its actual value.

Let us consider a null hypothesis that is a (sharp) simple statistical hypothesis H. This is often done in statistical practice although it would usually be more realistic to lump in with H a small 'neighborhood' of close hypotheses as in Good (1950, p. 97). The non-null hypothesis $\bar{H}$ is usually a logical disjunction of a continuous infinity of simple statistical hypotheses which I shall call the *components* of $\bar{H}$. These components will be at various 'distances' from H and some of them will be very close to it. Hence it is often sensible to talk of testing H *within* $\bar{H}$ although statisticians often use the word 'against'. An experiment or observation usually consists of a sample and when we refer to the size of an experiment we mean the size of the sample. (Neyman calls the α-level of a test the size of the test but I prefer to call it the α-level. We define α-levels later.) To distinguish H from the close components of $\bar{H}$ requires a large experiment. If no prior distribution is assigned to the components of $\bar{H}$ (given $\bar{H}$), then one could never obtain evidence in favor of H even if it is true, because there will be components of $\bar{H}$ close enough to H to be indistinguishable from it given any sample of a specified size. If, however, H is false, then a large enough sample should demonstrate its falsity to a Fisherian. On the other hand if a prior over the components of $\bar{H}$ is assumed, then a Bayesian or Doogian can obtain evidence in favor of H, though typically this evidence is not very great. Of course if H is redefined to be a composite hypothesis by including within H a small 'neighborhood' of the 'sharp' null hypothesis, then it becomes possible to obtain much more evidence in favor of H even without assuming a prior over the components of $\bar{H}$. Similarly if by the truth of Newtonian mechanics we mean that it is approximately true in some appropriate well defined sense we could obtain strong evidence that it is true; but if we mean by its truth that it is exactly true then it has already been refuted.

Very often a statististician doesn't bother to make it quite clear whether his null hypothesis is intended to be sharp or only approximately sharp. He also often has in mind an experiment of moderate size but does not state this explicitly. It is hardly surprising then that many Fisherians (and Popperians) say that "you can't get [much] evidence in favor of a null hypothesis but can only refute it." Regarding this statement itself as a kind of null hypothesis (!) the Fisherian's experience tends to support it (as an approximation) Doogianwise so the Fisherian (and Popperian) comes to believe it (because he is to some extent a Doogian without knowing it).

Perhaps the simplest example of a significance test is a test for whether a binomial physical probability p is equal to ½. I am assuming the sample to be definitely binomial and that only the value of p is uncertain. Suppose there are r 'successes' in n 'trials' so that the 'sigmage' is $x = (r - \frac{1}{2}n)/(\frac{1}{2}\sqrt{n})$. The larger is $|x|$, the more the sample differs from the most probable result (given the null hypothesis H).

In conversation I have emphasized to other statisticians, starting in 1950, that, in virtue of the 'law of the iterated logarithm', by optional stopping an arbitrarily high sigmage, and therefore an arbitrarily small tail-area probability, can be attained even when the null hypothesis is true. In other words if a Fisherian is prepared to use optional stopping (which usually he is not) he can be sure of rejecting a true null hypothesis provided that he is prepared to go on sampling for a long time. The way I usually express this 'paradox' is that a Fisherian can cheat by pretending he has a train to catch like a gambler who leaves the table when he is ahead. Feller (1950, p. 140) discusses optional stopping in a 'fair' gambling game and points out that, in virtue of the law of the iterated logarithm, an infinitely rich gambler can be sure of winning if he has the right to stop playing when he chooses (much good it would do him). Surprisingly, Feller does not mention the effect that optional stopping would have on Fisherian tail-areas. The point is implicit in Jeffreys (1939, Appendix I) and explicit in Good (1956a) and in Lindley (1957).

It is intuitively obvious in the binomial sampling experiment that the evidence is summarized by the values of r and n and that there is no point in investigating whether there really was a train to catch except as a criterion regarding the honesty of the statistician.[4] The use of tail-area probabilities is therefore logically shaky but I believe it is useful all the same, and can often be given a rough Doogian justification, at least when samples are not very large.

It also follows from the likelihood principle alone, without the Bayesian theory, that optional stopping is irrelevant when judging whether $p = \frac{1}{2}$.[5]

What is 'more extreme'? The idea that one outcome is more extreme than another one depends on whether it seems to be 'further away' from the null hypothesis. What is meant by 'further away' depends on some ordering, precise or vague, of the components of $\bar{H}$. [For a discussion of such orderings see, for example Kempthorne and Folks (1971, pp. 226–230).] The statistic chosen for testing the null hypothesis is chosen to reflect this distance. Thus the statistic is a 'discrepancy measure' to use the terminology of, for example, Kalbfleisch and Sprott (1976, p. 264). Sometimes more than one statistic is used for the same data because the statistician or his client has more than one non-null hypothesis in mind, each non-null hypothesis being composite. For example a distribution might be tested as a fully specified normal distribution within the class of all normal distributions or within the class of all possible distributions, and different test criteria would be used in these two cases.

The representation of the idea that one outcome is more extreme than another depends on the statistic (= function of the observations) that is used.

For example, suppose that the frequencies in the cells of a multinomial distribution having t categories are $n_1, n_2, \ldots, n_t$ (where $\Sigma\, n_i = N$, the sample size) and that the null hypothesis H specifies that the cell 'physical probabilities' (= propensities) are $p_1, p_2, \ldots, p_t$, where $\Sigma\, p_i = 1$. (A mnemonic for the symbol t, as used here, is that it is often the initial letter of the number of categories such as two, three, ten, twenty, twenty-six, thirty or a thousand.) A statistic that is often used for testing H is X^2 or χ^2 defined by

$$X^2 = \Sigma \frac{(n_i - Np_i)^2}{Np_i} .$$

[X^2 is often called χ^2 (chi-squared) but χ^2 is also used for the asymptotic distribution of X^2 when $N \to \infty$ and when H is true, so the more modern notation for the statistic is X^2.] There are other statistics used for testing H: see, for example, Good (1967) and Good and Crook (1974). One special appeal of X^2 to the non-Bayesian is that it resembles Euclidean distance, so it has geometrical simplicity, and the resemblance is especially marked when the p's are all equal to $1/t$ (the 'equiprobable case'). Moreover the asymptotic distribution of X^2 (given H), when N is large, is known and tabulated. It is the chi-squared distribution, as we mentioned before, with $t - 1$ 'degrees of freedom'.

It is intuitively to be expected, and also provable, that if the true probabilities differ from $p_1, p_2, \ldots, p_t$, then X^2 is likely to be larger than if H is true, and roughly speaking the more they differ the larger X^2 is likely to be. Therefore it makes at least some sense to say that one sample is 'more extreme' than another one if the former has a larger value of X^2. If some other statistic is used then 'more extreme' will have a (somewhat) different meaning. In other words the choice of a 'criterion' (= statistic) for testing a hypothesis H always involves some concept of what alternatives to H are under consideration, however vague or precise those alternatives might be, and the deviation of the statistic from its null value should correspond roughly to greater distances from H of the components of $\bar{H}$, the distances being defined in 'parameter space' (or more generally in hypothesis space). Of course 'distance' is here to be interpreted in a generalized sense, and it need not be a quadratic function of the parameters. One appealing distance function for multinomials is the 'expected weight of evidence per observation', $\Sigma\, p_j \log (p_j/q_j)$, for discriminating the null hypothesis from some other hypothesis. (The continuous analogue is obvious.) This suggests the use of the statistic $\Sigma\, p_j \log (tp_j/n_j)$ or of $\Sigma\, n_j \log [n_j/(tp_j)]$. The latter statistic comes to essentially the same thing as the familiar Neyman–Pearson–Wilks likelihood-ratio statistic. The likelihood-ratio statistic for testing a hypothesis H 'within' a hypothesis H' (where H' at least is composite) is defined as $\max_i P(E|H_i)/\max_j P(E|H'_j)$, where (i) E is the observed event; (ii) H is the disjunction of all H_i, (iii) H' is the disjunction of all H'_j, and (iv) H_i and H'_j are simple statistical hypotheses for all i and j. By saying that H is 'within' H' we mean that each H_i is a component of H' (besides being a component of H), though the converse is of course not true. Clearly the likelihood ratio tends to be small if H is false but H' is true, and tends to be smaller the farther away is the true H'_j from any of the H_i's. In this way Neyman and Pearson produced a technique of some generality for generating reasonable test criteria, framed in terms of the non-null hypothesis of interest to the statistician. Moreover Wilks found, in terms of X^2, the asymptotic distribution of the likelihood ratio for large sample sizes, under somewhat general conditions, and this made the likelihood ratio convenient to use in the Fisherian manner.

Notice how the likelihood ratio is analogous to a Bayes factor which would be defined as a ratio of weighted averages of $P(E|H_i)$ and of $P(E|H'_j)$. The weights would be $P(H_i|H)$ and $P(H'_j|H')$ but these probabilities are *verboten* by the non-Bayesian. Just as the value of an integral is often very roughly monotonically related to the maximum value of the integral so one hopes that the Bayesian averages are roughly related to the maximum likelihoods. It would not be entirely misleading to say that the use of the Neyman–Pearson–Wilks likelihood ratio is the non-Bayesian's way of paying homage to a Bayesian procedure. (I have expressed this argument incompletely to save space.)

Although the use of tail-area probabilities may reasonably be called 'Fisherian tests of significance' because Fisher made so much use of them, and developed their theory to such an extent, they had a long previous history. One early use was by Daniel Bernoulli (1734). [This example is described by Todhunter (1865, p. 222) and was used by Barnard (1979), in a Bayesian conference in Spain, as a challenge to Bayesians to explain why it was reasonable in Bayesian terms. I was able to reply only briefly at that conference, mainly by giving references to my own work, as usual, where this kind of question had been discussed.]

In Bernoulli's example the null hypothesis is that the normals to the planes of rotation of the various planets are uniformly distributed in direction. One of the statistics he used was the maximum angle between the planes of two planetary orbits. (Todhunter points out some improvements to Bernoulli's approach, but these improvements are barely relevant to the present philosophical discussion.) Clearly this statistic will tend to be small if all the planets rotate approximately in a single plane. Thus the appropriate meaning for 'more extreme' in this case is 'smaller'. Note that the statistic selected by Bernoulli was presumably chosen by him because he had both a null hypothesis and a vague non-null hypothesis in mind. The null hypothesis was that the normals were uniformly distributed in direction, whereas the vague non-null hypothesis was that the normals had a tendency to be close together. He wouldn't have tested the null hypothesis by looking at the maximum angle between planes of orbits unless he had in mind the (vague) non-null hypothesis just described. The smaller the maximum angle, the further away is the specific form of non-null hypothesis likely to be (away, that is, from the null hypothesis) and, at the same time, the less probable is the null hypothesis. I believe it is a feature of all sensible significance-test criteria that they are chosen with either a precise or (more often) a vague non-null hypothesis (usually composite) in mind. In this respect 'non-Bayesians' act somewhat like Bayesians. If a tail-area probability is small enough then it is worth while to try to make the non-null hypothesis less vague or even to make it precise, and the smaller the tail-area probability, the more worth while it is to make this attempt.

The 'tail-area probability' that Bernoulli obtained would have been somewhat less impressive if he had been able to allow for Pluto because its orbit makes a large angle with the other orbits. To allow for the possibility of such an 'outlier' a different statistic could be used, such as the average angle between all pairs of orbits instead of the maximum angle. Or, a little more artificially, one might use the maximum angle, allowing the deletion of one planet, but then some 'payment' would be required to allow for the artificiality. It is intuitively obvious that if enough of the normals to the orbits are close enough together, then there must be a physical reason for it even if some of the normals have a large deviation. The method of deleting one of the planets provides an example of selecting a significance test after looking at the data. Many textbooks forbid this. Personally I think that Rule 1 in data analysis is 'look at the data'.

The question of whether the Bode–Titius Law is 'causal', a subject that was treated by Good (1969) and by Efron (1971), is similar to Bernoulli's problem. It is noteworthy that one of the discussants of the former of these papers used the exceptional planets, combined with a t-test, to argue that "we are in the classic situation which Herbert Spencer once called a 'scientific tragedy' – a theory killed by a fact." Having failed to understand that Spencer's remark does not apply to the question of whether the Bode–Titius Law needs an explanation, he repeated the same fallacy in the discussion of

Efron's paper and was again refuted. That a reputable statistician could make such an error, and then to persist in it, shows the dangers of being misled by standard cook-book recipes when the conditions for their application are not present. In other words it is an example of the 'tyranny of words'. A misconception that sometimes arises from the same tyranny is the belief that a *P*-value less than 5% means that the null hypothesis has probability less than 5%! An example of this blunder is mentioned in a book review in *Applied Statistics* 28 (1979), p. 179.

Another example where a standard technique is inadequate was mentioned by Good (1956b, p. 253).

A sample of 100 readings is taken from some distribution for which the null hypothesis is that the readings are independently distributed with a normal distribution of zero mean and unit variance. It is decided in advance of sampling to divide the normal distribution up into ten equal areas, and to apply the χ^2 test to the ten-category equiprobable multinomial distribution of frequencies with which the readings fall into the ten areas. This would appear to be a very reasonable statistic. But what if it leads to a non-significant result even though one of the 100 readings was 20 standard deviations above the mean?

This example again shows that it is sometimes sensible to decide on a significance test after looking at a sample. As I've said elsewhere this practice is dangerous, useful, and often done. It is especially useful in cryptanalysis, but one needs good detached judgment to estimate the initial probability of a hypothesis that is suggested by the data. Cryptanalysts even invented a special name for a very far-fetched hypothesis formulated or pattern noted after looking at the data, namely a 'kinkus' (plural: 'kinkera'). (For a more accurate definition of a kinkus see Good (1980).) I agree with Herman Rubin's remark, at the Waterloo conference on scientific inference in 1970, that "A good Bayesian does better than a non-Bayesian but a bad Bayesian gets clobbered." Fisher once said privately that many of his clients were not especially intelligent, and this might have been part of his reason for avoiding Bayesian methods.

A very common example, to support the view that Fisherians allow for the non-null hypothesis, is that they often have to choose whether a tail-area probability should be single-tailed or double-tailed. If we are considering two treatments, and the null hypothesis is that there is no (or a negligible) difference in their effects (the null hypothesis) then it will usually be appropriate to use a double-tail; but, if the question (= the non-null hypothesis) is only whether the second treatment is *better* than the first, then a single-tail is usually appropriate. Clearly a sensible criterion should embody what we want to mean by 'more extreme' in terms of the components of the non-null hypothesis. We shall have more to say later regarding the choice of a significance-test criterion, in relation to the 'Bayes/non-Bayes synthesis' and to 'surprise indexes'. It is curious that Fisher introduced general features of statistics for estimation purposes, but not for significance tests. He seemed to select his significance-test criteria by common sense unaided by explicit general principles. Some such general principles were later introduced by Neyman and Pearson who made use of the 'power function' of a significance test and of the likelihood-ratio test with its attractive amount of generality.

It is sometimes stated by Fisherians that the only hypothesis under consideration is the null hypothesis, but I am convinced that this is only a way of saying that the non-null hypothesis is vague, not that it does not exist at all. As so often, in other contexts, the controversy (of whether there is always a non-null hypothesis) is resolved by saying "It depends what you mean". It is a common fallacy that if a concept is not precise then it does not exist at all. I call this the 'precision fallacy'. If it were true then it is doubtful whether any concepts would have any validity because language is not entirely precise though it is often clear enough. It is largely because of the precision fallacy that many statisticians are not Doogians, or do not know that they are. They think that Bayesians all use precise probabilities and do not recognize that the Doogian, who is in many respects a Bayesian, in principle uses 'upper and lower probabilities'.

To show that tail-area probabilities do not contain all that is needed we ask what if there had been a billion planets and the tail-area of Bernoulli's statistic had been 0.001? Would we then have rejected the null hypothesis? Surely we would have been surprised that the tail-area probability was not much smaller than 0.001, given the non-null hypothesis. Similarly, in the multinomial example, if N were exceedingly large, you would expect to get a very small value for $P(\tilde{X}^{2} \geqslant X^2)$ if the null hypothesis were false (where the tilde converts X^2 into a random variable). In other words, there are values of X^2 that are surprising whether the null hypothesis or the non-null hypothesis is true. Jeffreys (1939, p. 324) says that in such circumstances both he and Fisher should be very doubtful. The following Bayesian approach makes the matter clearer.

Bayes factors and tail-area probabilities. If a criterion S (a statistic) is sensibly chosen it might exhaust most of the weight of evidence against (or for) a null hypothesis H in the sense that $W(H:S) = W(H:E)$. For reasons to be discussed later, X^2 is such a statistic when testing the multinomial null hypothesis mentioned before. Now the asymptotic distribution of X^2 (given

H) when $N \to \infty$, is known. As I said before, it happens to be the χ^2 distribution with $t - 1$ degrees of freedom, though it is not essential that the reader should know what this means. This asymptotic distribution is fairly accurate, up to tail-areas as small as 1/100 at any rate, even if N is as small as t. (Some textbooks say that N needs to be as large as $5t$ but it has been known for some time that this is unnecessary.) Suppose we can make some guess at the distribution of X^2 given the (vague) non-null hypothesis, for a given value of N. It seems reasonable to me to assume that it will often resemble a Pearson Type 3 distribution, but with an extra thick tail on the right, and one way of guessing it is to specify its median, quartiles and other quantiles, depending on the application and on 'your' judgment. (The quantiles can be judged by imagining bets at various appropriate odds.) The density curves given the null and (vague) non-null hypothesis might have the general appearance shown in Figure 1.

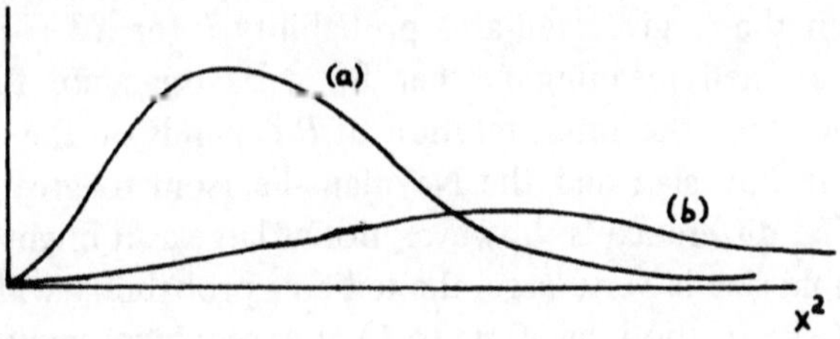

Fig. 1

Fig. 1. Probability densities of X^2 given H (curve a) and given $\bar{H}$ (curve b).

When N increases, curve (a), which represents the density of X^2 given H, hardly changes, but the mode of the curve for $\bar{H}$ (curve (b)) moves to the right. Now the Bayes factor against H, provided by X^2, is the ratio of the ordinate of curve (b) to that of curve (a) at the value of X^2 that occurs. Thus, as N increases, a given value of X^2 provides less and less evidence against H, indeed it ultimately provides an increasing weight of evidence in favor of H; and the same is true of a given tail-area probability $P(\tilde{X}^{-2} \geqslant X^2)$.

The above argument, showing that a fixed tail-area probability has diminishing evidential value against H as the sample size N increases, can also be expressed in terms of 'errors of the first and second kinds' (as defined below when we discuss the Neyman–Pearson approach). For we can concentrate our attention on any one, say H_j', of the simple statistical hypotheses of which the non-null hypothesis is a logical disjunction. We can then replace the curve (b) by the probability density curve of H_j'. For any fixed α, the value of β corresponding to H_j' increases as N increases and tends to 1 as $N \to \infty$. Moreover the simple likelihood ratio, for comparing H with H_j', which is a ratio of ordinates (instead of being a ratio of areas), tends to infinity as $N \to \infty$ if α is fixed. Thus, for any fixed α, however small, the evidence *in favor* of the null hypothesis tends to infinity as $N \to \infty$. Since this is true for each H_j' it must also be true when H is contrasted with $\bar{H}$. This result hardly requires a Bayesian argument to support it, but only the concept that the simple likehood ratio is monotonically related to the strength of the evidence.

Another way of looking at the above argument is that, for a fixed α, as N increases, more and more of the components of $\bar{H}$ become effectively refuted, and we are left with a contracting set of components in a smaller and smaller neighborhood of H. Thus H becomes more and more probable and this result does not even depend on our regarding H as itself only an approximation. Of course the argument is stronger still if we do regard H in this light.

We see then that a given tail-area probability P for X^2 (or for any other statistic) has a fixed meaning neither for a Bayesian nor for a Neyman–Pearsonian, and that the interpretation of P depends on the sample size. In this respect the Bayesian and the Neyman–Pearsonian agree to differ from a Fisherian. The difference is, however, not as large as it might seem because, when the sample size is very large, the tail-area probability will usually either be moderate (e.g., in the range 0.05 to 1) or exceedingly small (e.g., less than 1/10,000), depending on whether the null hypothesis is true or false.

These comments make one uneasy about certain experiments in parapsychology where the experiments are large and the 'proportional bulges' are small. For example, the best subject in Helmut Schmidt's experiments, in a binomial experiment, had a success rate of 52.5% in 6400 trials. (See Frazier, 1979.) As good a result would occur about 1/30 000 of the time in a sample of this size, if $p = \frac{1}{2}$. But, if we assume that, under the non-null hypothesis, p is uniformly distributed from 0.5 to 0.55 then the Bayes factor against the null hypothesis from this sample is only about $(80\pi)^{\frac{1}{2}}e^8 \div 320 \approx 150$. The factor could be larger if the prior density, given $\bar{H}$, tended to infinity near $p = \frac{1}{2}$, a possibility suggested by Good (1962b, pp. 74 and 78), where the matter is discussed in somewhat more detail.[6]

In spite of my logical criticisms of the use of tail-area probabilities, I have found that, for sample sizes that are not extremely large, there is usually an approximate relationship between a Bayes factor F and a tail-area probability P. I shall now discuss this relationship.

Good (1950, pp. 93–95) made use of the idea of basing a Bayes factor on the values of a statistic (not necessarily a sufficient statistic), such as X^2. This requires a judgment of the distribution of the statistic given $\bar{H}$. Given H, the distribution of X^2 is often uncontroversially known approximately. I said that "It would often happen that the factor in favour of H [the null hypothesis] obtained in some such way [by assuming a distribution for a statistic both given H and given $\bar{H}$] would be in the region of three or four times $P(X_0^2)$ [the tail-area probability of the observed value of the statistic given H]." In a footnote I remarked that "There are two independent reasons why the factor in favour of H exceeds $P(X_0^2)$. The first is that to pretend that the result is $X \geqslant X_0$ when it is really $X = X_0$ is unfair to H. The second is that $P(X \geqslant X_0 \mid H) < 1$, so that the factor from the evidence '$X \geqslant X_0$' is

$$P(X \geqslant X_0 \mid H)/P(X \geqslant X_0 \mid \bar{H}) \geqslant P(X \geqslant X_0 \mid H) = P(X_0^2)."$$

Good (1957; based on lectures of 1955 in Princeton and Chicago) took the matter somewhat further in the following words:

The Bayes/non-Bayes synthesis is the following technique for synthesizing subjective and objective methods in statistics. (i) We use the neo/Bayes–Laplace philosophy in order to arrive at a factor, F, in favour of the non-null hypothesis. For the particular case of discrimination between two simple statistical hypotheses, the factor in favour of a hypothesis is equal to the likelihood ratio, but not in general. The neo/Bayes–Laplace philosophy usually works with inequalities between probabilities, but for definiteness we here assume that the initial distributions are taken as precise, though not necessarily uniform. (ii) We then use F as a statistic and try to obtain its distribution on the null hypothesis, and work with its tail-area probability, P. (iii) Finally we look to see if F lies in the range

$$\left(\frac{1}{30P}, \frac{3}{10P}\right).$$

If it does not lie in this range we think again.

(Note that F is here the factor *against* H.)

In Good (1967) I examined numerical evidence relating a Bayes factor F against H to an orthodox tail-area probability. The application was to significance tests for equiprobability in a multinomial distribution and the Bayes factor was based on all the evidence. It turned out that, in 18 samples,

$$\frac{1}{6P\sqrt{(2\pi N)}} < F < \frac{6}{P\sqrt{(2\pi N)}},$$

where N is the sample size, and in 14 of the samples the 6 could be replaced by 3. The reason for introducing $N^{1/2}$ into the formula was its occurrence in Jeffreys (1939, Appendix I). Jeffreys showed in effect that, in a number of examples, the Bayes factor against the null hypothesis is roughly inversely proportional to $N^{1/2}$ for a given tail-area probability.

For the same multinomial problem, but with a much larger collection of samples, Good and Crook (1974) found that $P(F)$, the tail-area probability of F itself, was proportional to $(FN^{1/2})^{-1}$ within a factor of 5. For most pairs (t, N), where t designates the number of multinomial categories, we found that $P(F)$ is almost equal to a mathematical function of t, N, and F and does not otherwise depend much on the frequency count $(n_1, n_2, \ldots, n_t)$. This confirmed a basic idea of the Bayes–Fisher compromise, that a *Bayes factor is roughly some function of its own tail-area probability.*

Some similar numerical results were obtained for contingency tables in Crook and Good (1979).

In virtue of these results, and in spite of my propaganda for the Jeffreys–Good–Lindley paradox, I personally am in favor of a Bayes/non-Bayes compromise or synthesis. Partly for the sake of communication with other statisticians who are in the habit of using tail-area probabilities, I believe it is often convenient to use them especially when it is difficult to estimate a Bayes factor. But caution should be expressed when the samples are very large if the tail-area probability is not extremely small.

The Neyman–Pearson approach. The basic idea in the Neyman–Pearson approach to significance tests (Neyman and Pearson, 1933) is to consider the probabilities of errors of the first and second kinds. An error of the first kind is defined as the rejection of the 'null hypothesis' H when it is true, and an error of the second kind is the acceptance of H when it is false. The probabilities of these two kinds of error depend of course on the significance test that is to be used, and the two probabilities are usually denoted by α and β. Here β is regarded as a function of the components of $\bar{H}$ and $1 - \beta$ is called the *power function*. Also α might be a function of the components of H if H is composite, but we shall suppose for simplicity that H is simple. If the significance test is defined in terms of a statistic with a threshold, then α depends on the threshold. Sometimes the threshold is chosen so that α takes some conventional value such as 0.05; otherwise the power function depends on the value of α.

The concept of a power function is in my opinion most useful when $\bar{H}$ is a disjunction of a single-parameter set of components. When there is more than one parameter the power function is liable to be difficult to apprehend.

It is not by any means always sensible either to accept or to reject a hypothesis in any sharp sense. If a P-value is very small this gives more reason to look for sharp alternatives to it, in a Fisherian context, and this important fact is not captured by the Neyman–Pearson technique.

Neyman and Pearson recognized that a hypothesis cannot be tested unless there is some notion of alternatives so they formalized the matter by assuming the alternatives could be specified accurately. In this respect they were going in the direction of the 'sharp' Bayesian, but they stopped short of assuming a conditional prior for the components of $\bar{H}$ (conditional on $\bar{H}$).

That the notion of errors of the second kind is useful for orthodox statistics, in one-parameter problems, whatever its weaknesses may be, can be seen from Frieman *et al.* (1978). They found that the emphasis on α (or P-value) and the neglect of β had led to many ineffective clinical trials. In 71 'negative' randomized control trials, chosen in a sensible manner from published papers, 50 of the trials had a 10% risk of missing a 50% therapeutic improvement (in a well defined sense). This poor performance might have been avoided if the experimenters had allowed for errors of the second kind when planning their experiments. They would have realized that their samples were too small. But the smallness of the samples in these trials was presumably caused also by the ethical consideration, that as soon as one treatment seems better than another it seems unethical to use the apparently less good treatment. (See also Good (1978), which gives a brief recent discussion of ethical problems in clinical trials. It seems that the ethical difficulty can only be overcome either by social contract or if patients voluntarily accept compensation for entering the trial.[7])

It is of historical interest that Neyman (1977) says that he and Pearson were inspired to formulate their theories by Borel (1920) who "insisted that: (a) the criterion to test a hypothesis (a 'statistical hypothesis') using some observations must be selected *not after the examination of the results of observation*, but before, and (b) this criterion should be a function of the observations 'en quelque sorte remarquable'."

When using the Neyman–Pearson theory, at least in its standard form, the precise significance test and the value of α are supposed to be determined in advance of the experiment or observation. I have already argued that one cannot always sensibly determine a significance test in advance, because, heretical though it may be in some quarters, sometimes the data overwhelmingly suggests a sensible theory after the data are examined. On some other occasions the suggestion is not overwhelming and then it is desirable to collect more data if this is not too expensive. In Daniel Bernoulli's example, and in the example of the Titius–Bode Law, and in many cryptanalytic problems, it is difficult or impossible to collect more data, and in some other situations it is very expensive to do so.[8]

On some occasions the non-null hypothesis is of high dimensionality, or is otherwise a very complicated disjunction of simple statistical hypotheses, with the result that the statistician is unable to apprehend the power function intuitively. In such cases it may be possible to average the power function with weights that are proportional to the prior probabilities so as to obtain a convenient summary of the power function. (Good and Crook, 1974, p. 711.) To do this can be regarded as an example of a Bayes/Neyman–Pearson compromise. Some numerical examples of this procedure are given by Good and Crook (1979).

Surprise. The evolutionary value of surprise is that it causes us to check our assumptions (Good, 1956c, p. 1131). Hence if an experiment gives rise to a surprising result, given some null hypothesis H, it might cause us to wonder whether H is true even in the absence of a vague alternative to H. It is therefore natural to consider whether a statistical test of H might be made to depend upon some 'index of surprise'.

There is one such index due to Warren Weaver (1948), and a generalization due to Good (1954, 1956c). Weaver proposed the index $(\Sigma p_i^2)/p$ where p_i runs through the probabilities of mutually exclusive possible outcomes of an experiment (or observation) and p denotes the probability of the event that actually occured. This index can be written in the form

$$\lambda_1 = \mathscr{E}(\tilde{p})/p$$

where $\tilde{p}$ denotes the random variable whose possible values are the probabilities of outcomes and $\mathscr{E}$ denotes mathematical expectation. [When there is a continuous infinity of possible outcomes one can of course use probability densities in place of probabilities, but the index is not invariant under transformations of the independent variable. It is invariant under linear transformations.] This form for λ_1 shows that it can replace the microscopic probability of the outcome by a number of more reasonable magnitude. For example, if part of the outcome of spinning a coin is that it lands on a certain point on the floor (which we regard as irrelevant to whether the coin is biased) the microscopic probability of this irrelevance will cancel out when the ratio λ_1 is computed. The surprise index depends on how the outcomes are grouped, especially when there are a continuous infinity of possible outcomes.

As a generalization of λ_1, Good (1954, 1956c) proposed

$$\lambda_u = (\mathscr{E} p^u)^{1/u}/p \qquad (u > 0),$$

and the limiting form ($u \to 0$)

$$\lambda_0 = \exp\{\mathscr{E}(\log \tilde{p}) - \log p\}$$

which is the ratio of the 'geometrical expectation' of $\tilde{p}$ to p. If we write $\Lambda_u = \log \lambda_u$, then Λ_u is additive if the results of several statistically independent experiments are regarded as a single experiment whereas the indexes λ_u are multiplicative.[9] An additive property seems more natural. In particular

$$\Lambda_0 = \mathscr{E}(\log \tilde{p}) - \log p$$

and this can be interpreted as the amount of information in the event that occurs minus the expected amount. (See Good, 1950, p. 75.) The most natural of the surprise indexes[10] are λ_1, Λ_1, λ_0 and Λ_0. Moreover Λ_0 has the desirable property that its expectation is zero, that is, the expected amount of surprise vanishes if the null hypothesis is true. (Compare Bartlett (1952).)

If we prefer not to assume anything about the non-null hypothesis, then a surprise index can sometimes reasonably be used for deciding whether the null hypothesis is suspect, or whether it is supported (as when Λ_0 is negative). Surprise indexes are not yet in common use in statistical practice, but perhaps they should be. One could also make a surprise/Fisher compromise by using the tail-area probability of a surprise index as a criterion for testing a null hypothesis. But sometimes a surprising event is regarded as 'merely a coincidence' because we cannot think of any reasonable alternative to the null hypothesis.

I have here ignored some problems of definition of surprise that are related to the meaning of 'simplicity'. The matters are discussed to some extent by Good (1956c).

The Bayesian influence on significance testing. The coherence of the Bayesian philosophy, especially in its Doogian form, can shed light upon and can influence non-Bayesian procedures.

For example, Bochner (1955) once asked me "Why use X^2 [for testing a multinomial null hypothesis H] ?" My reply was that, under a wide variety of Bayesian models, the weight of evidence against H is approximately a linear function of X^2 when the weight of evidence is not large. This provides a rough justification for combining independent X^2's by taking a linear combination of them (becuase independent weights of evidence are additive).

Another example is a proposed procedure for combining several tail-area probabilities P_1, P_2, P_3, ... that are obtained by distinct significance tests based upon the same data. An informal Doogian argument leads to the rule of thumb of regarding the harmonic mean of $P_1, P_2, P_3, \ldots$ as a resultant tail-area probability provided that nothing better comes to mind (Good, 1958).

4. EXPLICATIVITY

So far in this article the emphasis has been on whether a hypothesis is probable, but the selection of a hypothesis depends also on its utility or on a quasi-utility such as its power to predict or to explain. If this were not so we would always prefer a tautology such as 2 = 2 to any more informative hypothesis. An analysis that allows especially for explanatory power is given by Good (1968, 1977). The analysis introduces a concept called 'explicativity' which leads to a sharpened form of the Ockham–Duns razor and which is found to shed light both on significance tests and on problems of estimation.

Virginia Polytechnic Institute and State University

NOTES

* University Distinguished Professor of Statistics, Virginia Polytechnic Institute and State University, Blacksburg, Virginia, 24061. This work was supported in part by the National Institute of Health, Grant No. 18770.

[1] I am in the habit of calling non-Bayesian methods 'orthodox' because for some years after World War II I was almost the only person at meetings of the Royal Statistical Society to defend the use of Bayesian ideas. Since such ideas are now much more popular it might be better to call non-Bayesian methods 'sampling theory' methods and this is often done in current statistical publications. This name covers the use of tail-area probabilities ('Fisherian') and probabilities of errors of the first and second kinds (Neyman–Pearsonian).

[2] Intensities of conviction are often called 'degrees of belief' but in 1946 or 1947 Gilbert Ryle rejected an article of mine, for *Mind* (containing an outline of Good, 1950), partly on the grounds that you either believe something or you do not (and also because the article was somewhat mathematical). He also described the interpretation of $P(A \mid B)$ in terms of degrees of belief, when B does not obtain, as 'make belief'. If I had used the expression "intensity of conviction" this joke would have been unavailable to him.

[3] By the time a pure likelihood man recognizes that initial probabilities must be allowed for, he calls them initial likelihoods or initial supports instead so as to avoid admitting that he has become a Bayesian!

[4] It is intuitively obvious to a layman if not to a Lehmann. Lehmann (1959) made a notable contribution to the sampling-theory approach to significance tests. My wisecrack

is not really fair to Lehmann because he recognizes the value of Bayesian decision theory, but there is no reference to optional stopping nor to the law of the iterated logarithm in the index of his book, so presumably his book does not mention the danger of optional stopping in a Fisherian context.

[5] It might be thought that binomial sampling, with fixed sample size n, leading to r 'successes', would lead to a different tail-area probability, for testing the hypothesis of a specified value for p, than if 'inverse binomial sampling' were used (in which r is fixed and sampling is done until r successes are achieved). It turns out, curiously enough, that, if r and n are unchanged, then the tail-area probabilities are equal for the two methods of sampling. This can be inferred from Patil (1960, p. 502). But the unbiased estimates for p are unequal, being r/n for binomial sampling and $(r - 1)/(n - 1)$ for inverse binomial sampling. (See, for example, Kendall and Stuart, 1960, p. 593.)

[6] Perhaps an acceptable prior, given $\bar{H}$, would be a mixture of beta distributions scaled to lie in the interval (½, 1). This prior would be similar to the one used in the hierarchical Bayes technique (Good and Crook, 1974). The implications for Schmidt's data could be worked out when the data are available.

[7] Gordon Tullock (1979) suggests that the volunteer might receive a conditional reward in a 'double-blind' experiment; that is, the reward would depend on whether the treatment was later revealed to be the more favorable or the less favorable one. It would also sometimes be possible to determine the compensation depending on the effectiveness of the treatment for the particular patient.

[8] In both the astronomical examples there is collateral information from the satellites of the planets. But in Bernoulli's problem the evidence was already overwhelmingly strong.

[9] The expression $\Lambda_u + \log p$ is sometimes called Rényi's generalized entropy (Rényi, 1961) but he was anticipated in 1954, as explained in the text.

[10] The indexes Λ_1 and Λ_0 are closely related to the quadratic and logarithmic fees that encourage honesty and 'objectivity of judgments'. These were defined respectively and independently by Brier (1950) and by Good (1952) and there is now a large literature on the topic. See, for example, Good (1972), Savage (1971), and Hendrickson and Buehler (1972).

REFERENCES

Barnard, G. A., 'On the Bayesian–antibayesian controversy', International Meeting on Bayesian Statistics, Valencia, Spain, 1979. To be published in *Trabajos de Estadística y de Investigacion Operativa.*

Bartlett, M. S., 'The statistical significance of odd bits of information', *Biometrika* 39 (1952), 228–237.

Bernoulli, D., 'Recherches physiques et astronomique . . . '. *Recueil des pièces qui ont remporté le prix de l'Academie Royale des Science* 3 (1734), 93–122.

Bochner, S., Private oral communication, 1955.

Borel, E., *Le Hasard*, Hermann, Paris, 1920.

Braithwaite, R. B., A lecture at the 1951 weekend conference of the Royal Statistical Society in Cambridge, England, 1951.

Brier, G. W., 'Verification of forecasts expressed in terms of probability', *Monthly Weather Rev.* 78 (1950), 1–3.

Crook, J. F. and Good, I. J., 'On the application of symmetric Dirichlet distributions and their mixtures to contingency tables, Part II', *Annals of Statistics* (in press) (1979).

Efron, B., 'Does an observed sequence of numbers follow a simple rule?', *J. Amer. Statist. Assoc.* 66 (1971), 552–568 (with discussion).

Feller, W., *An Introduction to Probability Theory and its Applications*, Vol. 1, Wiley, New York, 1950.

de Finetti, B., *Theory of Probability*, Vol. 1, Wiley, New York, 1975.

Frazier, K., 'Schmidt's airing at the APS', *The Skeptical Inquirer: The Zetetic* 3 (1979), No. 4, 2–4.

Frieman, J. A., Chalmers, T. C., Smith, Harry, Jr., and Kuebler, R. R., 'The importance of beta, the type II error and sample size in the design and interpretation of the randomized control trial', *New England J. Medicine* (1978), 690–694.

Good, I. J., *Probability and the Weighing of Evidence*, Charles Griffin, London; Hafners, New York, p. 119, 1950.

Good, I. J., 'Rational decisions', *J. Roy. Statist. Soc.* **B14** (1952), 107–114.

Good, I. J., 'The appropriate mathematical tools for describing and measuring uncertainty', Chapter 3 of *Uncertainty and Business Decisions*, Liverpool, second edition 1957, 20–36, 1954.

Good, I. J., Contribution to the discussion of a paper by G. S. Brown, in *Information Theory* (ed. C. Cherry), Butterworths, London, p. 13, 1956a.

Good, I. J., 'Which comes first, probability or statistics?', *J. Inst. Actuaries* 82 (1956b), 249–255.

Good, I. J., 'The surprise index for the multivariate normal distribution'. *Annals Math. Statist.* 27 (1956c), 1130–1135.

Good, I. J., 'Saddle-point methods for the multinomial distribution', *Annals Math. Statist.* 28 (1954), 861–881.

Good, I. J., 'Significance tests in parallel and in series, *J. Amer. Stat. Assn.* 53 (1958), 799–813.

Good, I. J., 'Subjective probability as the measure of a non-measurable set', *Logic, Methodology, and Philosophy of Science: Proc. of the 1960 International Congress*, Stanford University Press, pp. 319–329, 1962a.

Good, I. J., Contribution to the discussion in *The Foundations of Statistics*, opened by L. J. Savage, Methuen, London; Wiley, New York, 1962b.

Good, I. J., *The Estimation of Probabilities: An Essay on Modern Bayesian Methods*, M.I.T. Press, 1965.

Good, I. J., 'A Bayesian significance test for multinomial distributions', *J. Roy. Statist. Soc.* **B29** (1967), 399–431. (With discussion.)

Good, I. J., 'Corroboration, explanation, evolving probability, simplicity, and a sharpened razor', *Brit. J. Philos. Sci.* 19 (1968), 123–143.

Good, I. J., 'A subjective evaluation of Bode's Law and an "objective" test for approximate numerical rationality', *J. Amer. Statist. Assoc.* 64 (1969), 23–66.

Good, I. J., 'Twenty-seven principles of rationality', 1971, Appendix to 'The probabilitistic explication of information, evidence, surprise, causality, explanation. and utility', in V. P. Godambe and D. A. Sprott (eds.), *Foundations of Statistical Inference* (Proceedings of an international symposium at Waterloo, April 1979). Holt, Reinhart and Winston of Canada, Toronto, 1979, pp. 124–127.

Good, I. J., 'Information, rewards, and quasi-utilities', in J. J. Leach, R. Butts, and G. Pearce (eds.), *Science, Decision and Value*, D. Reidel, Dordrecht, 1973, 115–127.

Good, I. J., 'The Bayesian influence, or how to sweep subjectivism under the carpet', in *Foundations of Probability Theory, Statistical Inference, and Statistical Theories of Science* (Proc. of a Conference in May 1973 at the Univ. of W. Ontario; Eds. C. A. Hooker and W. Harper), Vol. 2, D. Reidel, Dordrecht, Holland, 1976, pp. 125–174.

Good, I. J., 'Explicativity: a mathematical theory of explanation with statistical applications', *Proc. Roy. Soc.* (*London*) **A354** (1977), 303–330. Reprinted in Zellner (1980).

Good, I. J., 'Ethical treatments', *J. Statist. Comput. Simul.* 7 (1978) 292–295.

Good, I. J., 'Some history of the hierarchical Bayesian methodology', invited paper for the International Meeting on Bayesian Statistics, May 28–June 2, 1979, Valencia, Spain. *Trabajos de Estadistica y de Investigacion Operativa* (in press).

Good, I. J., 'The philosophy of exploratory datum analysis'. In *American Statistical Association 1980 Proceedings of the Business and Economic Statistics Section* 1980.

Good, I. J., and Crook, J. F., 'The Bayes/non-Bayes compromise and the multinomial distribution', *J. Amer. Statist. Assoc.* 69 (1974), 711–720.

Good, I. J., and Crook, J. F., 'The Rank II powers of tests for multinomials and contingency tables', In preparation, 1979.

Hendrickson, A. and Buehler, R. J., 'Elicitation of subjective probabilities by sequential choices', *J. Amer. Statist. Assoc.* 67 (1972), 880–883.

Jeffreys, H., *Theory of Probability*, Clarendon Press, Oxford, 1939. (The third edn. appeared in 1961.)

Kalbfleisch, J. G. and Sprott, D. A., 'On tests of significance', in C. A. Hooker and W. Harper (eds.), *Foundations of Probability Theory, Statistical Theory, Statistical Inference, and Statistical Theories of Science, Vol. 2,* D. Reidel, Dordrecht, Holland, 259–272, 1976.

Kempthorne, O and Folks, L., *Probability, Statistics, and Data Analysis*, lowa State Univ. Press, 1971.

Kendall, M. G. and Stuart. A., *The Advanced Theory of Statistics*, Vol. 2, Charles Griffin, London, 1960.

Lehmann, E. L., *Testing Statistical Hypotheses*, Wiley, New York, 1959.

Lindley, D. V., 'A statistical paradox', *Biometrika* **44** (1957), 187–192.

Minsky, M. and Selfridge, O. G., 'Learning in random nets', in Colin Cherry (ed.), *Information Theory*, Butterworths, London, 1961, pp. 335–347.

Neyman, J., 'Frequentist probability and frequentist statistics', *Syntheses* **36** (1977), 97–131.

Neyman, J. and Pearson, E. S., 'On the problem of the most efficient tests of statistical hypotheses', *Philosophical Transactions of the Royal Society of London, Series A* **231** (1933), 289–337.

Patil, G. P., 'On the evaluation of the negative binomial distribution with examples', *Technometrics* **2** (1960), 501–505.

Peirce, C. S., 'The probability of induction', *Popular Science Monthly* (1878), reprinted in James R. Newman (ed.), *The World of Mathematics,* **2**, Simon and Schuster, New York, 1956, pp. 1341–1354.

Rényi, A., 'On measures of entropy and information', *Proc. Fourth Berkeley Sympos. Math. Statist. and Prob.*, Vol. 1, Univ. Press, Berkeley, Calif., pp. 547–561, 1961.

Savage, L. J., 'Elicitation of personal probabilities and expectations', *J. Amer. Statist. Assoc.* **66** (1971), 783–801.

Todhunter, I., *A History of the Mathematical Theory of Probability,* 1865. Reprint, Chelsea Publishing Co., New York, 1949 and 1965.

Tullock, G., Private oral communication (1979).

Turing, A. M., Private communication (1941).

Weaver, W., 'Probability, rarity, interest and surprise', *Scientific Monthly* **67** (1948), 390–392.

Zellner, A. (ed.), *Bayesian Analysis in Econometrics and Statistics: Essays in Honor of Harold Jeffreys.* North Holland, Amsterdam (1980).

DEBORAH MAYO

TESTING STATISTICAL TESTING*

1. PHILOSOPHY OF STATISTICS AND PHILOSOPHY OF ECONOMICS

At a recent conference on problems in economics the following problem was raised:

> Econometricians like to think of themselves as scientists, and their methods as scientific. Students of the philosophy of science, on the other hand, have not had any notable success in relating the formal concepts of scientific method or the logic of scientific explanation and theory construction to either the method or the theory of econometrics [3, p. 238].

This should not be taken to mean that the theory and practice of economics and econometrics is not scientific. It rather points to the need for a greater effort among philosophers to tie their analyses to actual scientific practice. More specifically, it indicates a need for philosophers of science to examine statistical theorizing in science, since inference and explanation in economics is often statistical in nature.

In building quantitative models of economic phenomena one may begin with an economic theory which is modelled mathematically, and then evaluate parameters of the model, or one may begin with a set of potentially relevant variables about which relationships are sought. In either case data must be gathered and values for the parameters of the model must be estimated or tested. The relevant consideration for our purposes is that whether it is due to the variability of the phenomena or the errors of observation and measurement, statistical methods are generally involved in evaluating this data. This raises questions about which statistical procedure to use and how its results bear upon the economic theory.

In the last 50 years or so a new battery of modern statistical techniques has become available, owing very heavily to the theories of Fisher [5] and Neyman and Pearson [21]. These techniques, often referred to as the *classical*, the *objective* or the *frequency* theory shall here be referred to as the *standard methodology* (SM). The application of statistics to economics gave birth to the science of econometrics. As with other scientists who applied statistics to their fields, workers in economics were more concerned with applying and absorbing statistical techniques into their science than with the foundational

J. C. Pitt (ed.), Philosophy in Economics, 175–203.

and philosophical questions associated with these methods. Perhaps this is as it should be for practitioners of science. It is, after all, the business of philosophers and not scientists to scrutinize the nature and epistemological basis of inference tools. With respect to statistical inference tools, this is the work of the philosophers of statistics.

Among those who were busy applying the SM, a growing controversy arose about which of the many new methods to use and how their results were to be interpreted. Several of those who searched for the valid basis for SM concluded that there was no valid basis. They claimed that the standard methods were misconceived and unable to perform the tasks for which they were being used in science. New inference philosophies arose which abandoned the principles of SM. Some were revivals of methods employed prior to modern developments, such as the Bayesian philosophy. This resulted in an even greater controversy over which methods to use. While the confusion and controversy has taken a number of turns, it is raging as strong as ever. As SM is increasingly being used in the sciences, it is increasingly coming under attack. With ever more vital decisions being based upon its methods the urgency in resolving the controversy is clear.

Given the close ties between economics and statistics, it is not surprising to find that problems concerning the evaluation of economic models are closely tied to controversies in the foundations and philosophy of statistics. By examining an important question in philosophy of statistics this paper hopes to be a contribution to the philosophy of economics.

The question upon which I shall focus is: What information is to be considered relevant for statistical inference? The particular sort of statistical inference to be focused on is *hypothesis testing*. The question has been such a major point of disagreement that the different testing philosophies may in general be distinguished on the basis of how they answer it. In particular, the standard and non-standard testing philosophies may be distinguished in this way.

The standard principles, that is, the principles of SM, deem relevant the manner in which data is generated and how various hypotheses to be evaluated by this data are specified. As a result, what would blithely be considered the same data and hypotheses may be treated differently according to how each was generated. In contrast, according to non-standard principles, in particular the likelihood principle, both the manner of generating data and specifying hypotheses are considered irrelevant for making inferences. It makes no difference to the result of a non-standard inference whether the hypothesis to be evaluated is specified *after the data* to be used in its evaluation is generated (SAD) or whether it is specified *before the data* (SBD). Thus, the non-standard theorist sees as unjustified any requirement which demands that hypotheses be specified in advance of the experiment. And so, the non-standard theorist in effect is demanding that the SM theorist justify his requirement for SBD. But, how can such a requirement be justified, based as it is on a view that appears to make the very time at which a hypothesis is formulated logically relevant to the evaluation of the hypothesis?

I shall attempt to provide such a justification by showing how the requirement permits SM to guarantee the reliability and precision (size and power) of its tests. But the standard criteria of reliability and precision are based upon the *sample space*, something considered irrelevant on non-standard approaches. As a result, such non-standard approaches will be shown to permit drawing faulty inferences with high probability – in extreme cases with probability one. As an example of a non-standard theory of testing I will employ the likelihood approach of Ian Hacking as set out in his *Logic of Statistical Inference* [11]. I next consider how the problem of permitting hypotheses to be SAD bears upon a Bayesian philosophy. It is concluded that non-standard theories of testing require additional principles concerning the generation of data and hypotheses if they are to avoid objectionable inferences. Whether there exists any non-standard justification for such principles is unclear.

Having shown how requiring hypotheses to be SBD enables SM to uphold its principles of reliability and precision, I argue that the rationale for using tests which are reliable and precise differs from the one given in the accounts of standard tests. Typically, the rationale given is that such tests make relatively few faulty inferences in the long run. The purpose of statistical tests in scientific contexts, as I see it, is to detect certain discrepancies between hypotheses or between data and hypotheses. In this way it provides information in order to assess substantive scientific hypotheses. The goal of low error rates in the long run is not viewed as an end in itself for tests. The importance of tests with 'good-error characteristics' is the ability to use them to formulate tests which can detect all and only discrepancies of a certain size. Hence by requiring hypotheses be SBD, and thereby ensuring certain error rates, one is ensuring the test is providing the information it is supposed to. This constitutes the basis for the objectivity of tests built upon SM.

2. BACKGROUND

A brief look at the background of the problem of inductive inference provides clues as to how non-standard inference theories seem to have arisen. What is sometimes referred to as the 'old' problem of induction arose from the empiricist principle whereby the acceptance and rejection of theories is to be solely the result of observation and experiment, and the recognition by Hume that no finite amount of experimentation can logically justify general theories. As a result, people began to consider whether favorable observations could be seen as providing a certain degree of proof or confirmation to theories. This gave rise to the 'new' problem of induction which has typically been seen as finding a quantitative relation *M* which is to measure the evidential strength between data *d* and a generalization or hypothesis *h*.

Though one could not be logically certain of hypothesis *h* given data *d*, the possibility that one could consider *h* as having some degree of probability suggested itself. Hence, it was natural to use probability functions in setting up the quantitative relation *M*. There have been a number of attempts to erect a theory of inductive inference by setting up an evidential relation *M*. Typically, *M* is defined as measuring degrees of belief, confirmation, support, credibility, plausibility and the like. All such attempts will be referred to as *evidential relationship* or E–R approaches.[1] Confronted with the same data and hypothesis an identical measure of evidential relationship is obtained by applying measure *M*. Thus, for E–R views, the manner of data generation and the order in which the hypotheses are specified are irrelevant to the measure *M* and hence irrelevant to inference.

It is from this conception of inference, that sees the evidential import of data to be found in a measure of E–R, that the Likelihood Principle apparently arose. As both the Bayesian and Likelihood philosophies incorporate some form of Likelihood Principle, both of these non-standard philosophies may be seen as step-children of Hume and the empiricists. Not only do these E–R views enjoy the neatness of being based upon a single quantitative measure *M*, they also have the simplicity of having a single inference principle typically exhorting one to maximize or increase the measure *M*. For instance, Hacking, following Carnapian logic, employs the likelihood function to measure the degree of strength given data affords a hypothesis. He then obtains a measure of the relative support afforded to rival hypotheses h_0 and h_1 by calculating the ratio of the likelihoods of each. It is the likelihood ratio that serves as his measure *M*. The fundamental inference principle is to reject hypotheses having rivals with much greater likelihoods. Before examining such non-standard theories, however, let us consider an example of a standard theory of testing.

3. STANDARD TESTING: NEYMAN–PEARSON (NP) TESTS

In order to apply the mathematical apparatus of statistics to the results of an experiment it must be construed in terms of a *statistical model*. That is, the experiment is viewed as the observation of a random variable D whose possible values d_i make up the sample space $\mathcal{X}$, and which has a certain probability distribution p_D. A statistical hypothesis is a statement about this distribution over $\mathcal{X}$. Typically, the general form of p_D is assumed and the hypothesis is an assertion about the value of one (or more) of the parameters θ governing the distribution.[2]

A function of D, $T(D)$, called the *test statistic* is specified. The test consists of a rule which designates, before the experiment is carried out, which of the values of $T(D)$ will be taken to reject the hypothesis under test, that is the *test* or *null hypothesis* h_0. The set of observations yielding these values of $T(D)$ make up the *critical region* of the test. The critical region is specified so that $T(D)$ will have a given probability, called the *size* or the *significance level* of the test, of falling into the region if the null hypothesis is true. This is just the probability of making what is called the *type I error*, that is, falsely rejecting h_0.

In the NP theory, an alternative hypothesis h_1 is also specified before the experiment, and it is to be accepted just in case h_0 is rejected. Then a second type of error, the *type II error* may be specified. A type II error is committed when h_0 is accepted (i.e., not rejected) when it is actually false and the alternative h_1 is true. The probabilities of making type I and type II errors are denoted by α and β respectively:

$$P(\text{reject } h_0/h_0 \text{ is true}) = \alpha$$
$$P(\text{accept } h_0/h_1 \text{ is true}) = \beta$$

3.1. *Size and Power*

The smaller the size α the more reliable the test, but it must be balanced with a reasonable β. If β is too large, the test is said to have insufficient *power* (or precision) where power is defined as $1 - \beta$. By simply never rejecting h_0, α can be reduced to 0. But then h_1 is never accepted, so β is 1 and hence the power is 0. Such a test would literally have zero power to detect the truth of the alternative. An *optimal* NP test having a given size α is one which at the same time maximizes the power (i.e., minimizes β). The experimenter is construed as first specifying α to be the maximum proportion of times he can 'afford' to make a type I error, and then maximizing the power. The context is construed as a pragmatic, decision-theoretic one, although as I shall later argue it need not be.

Optimal NP tests typically take the form: reject h_0 iff $T(D) \geqslant t$ for some fixed value t. The associated critical region is $\{d: T(d) \geqslant t\}$. This form arises because it is often reasonable to suppose h_0 should be rejected when the value of D is too far (in one or another direction) from the value (or values) of θ hypothesized by h_0. For example, in testing a hypothesis h_0 about the mean θ of a distribution where h_0 is $\theta = \theta_0$, it is reasonable to conclude h_0 is false if the observed or sample mean is too far away from the hypothesized mean. If random variable D is the sample mean, the test may take the form: reject h_0 just in case $|D - \theta_0| \geqslant \delta$ for some positive value of δ. As distance is conveniently measured in units of *standard deviations*[3] of the observable D (i.e., σ_D's), δ may be written as $k\sigma_D$. A test of this form is appropriate if one is interested in rejecting h_0 when the true value of θ is either less than or greater than θ_0. This is the case when dealing with the composite alternative hypothesis h_1 of form: $\theta \neq \theta_0$, and yields what is called a *two-sided* test.

If the alternative hypothesis states that θ has a value greater than θ_0, then one would want to reject h_0 only when positive differences between the observed and hypothesized mean were detected. The appropriate test would then take the form: reject h_0 if and only if $(D - \theta) \geqslant k\sigma_D$, for some positive k. This is an example of a *one-sided* test. By letting the test statistic $T(D)$ be $\frac{(D - \theta_0)}{\sigma D}$ the above one and two-sided tests become: reject h_0 iff $T(D) \geqslant k$ and reject h_0 iff $|T(D)| \geqslant k$, respectively. Subtracting from D its mean and dividing by its standard deviation yields the *normalization* of D. Hence, $T(D)$ is here the normalization of D.

The virtue of using the normalization of D as the test statistic $T(D)$ is that $T(D)$ is known to have (approximately) a *normal* distribution with a mean of 0 and a standard deviation of 1 (i.e., it is *standard normal*) regardless of the distribution of D. By selecting $T(D)$ to be a statistic whose distribution is known without having to know the value of any unknown parameter, (such as the one under test) it is possible to calculate the probabilities of the two types of errors prior to the experiment. In this way it can be ensured that these probabilities are sufficiently small regardless of which hypothesis is true. The distribution of $T(D)$ is called the *sampling distribution*, and it is the basis for both formulating and evaluating standard tests. Considering an example of a commonly used test may help to illuminate the standard testing concepts.

EXAMPLE (3–1): In an experiment to determine whether the probability that a coin lands heads is 0.5 or 0.8, 100 tosses are made and the proportion of heads observed. Let the experiment be modeled as the observation of random variable D with a *binomial distribution*[4] where D is the proportion of heads observed, θ the probability of heads, h_0 is $\theta = 0.5$ and h_1 is $\theta = 0.8$, and $T(D) = \frac{(D - 0.5)}{\sigma D}$. It is decided to reject h_0 iff the observed proportion differs from the hypothesized proportion by 2 or more standard deviation units. Then the test is:

$$\text{Reject } h_0 \text{ iff } T(D) \geqslant 2.$$

The size of this test is

$$P(T(D) \geqslant 2/\theta = 0.5) = \alpha$$

and it may be easily calculated to be approximately 0.03 since $T(D)$ has the standard normal distribution. (The probability that *any* standard normal variable is greater than or equal to 2 is 0.03 regardless of the value of θ.) Being the area to the right of 2, 0.03 is called the *tail area* of the distribution. As $\sigma_D = 0.05$ under the assumption that $\theta = 0.5$ the test is equivalent to the rule:

$$\text{Reject } h_0 \text{ iff } D \geqslant 0.6.$$

Then the probability of a type II error (i.e., accepting h_0 when the alternative is true) is

$$P(D < 0.6/\theta = 0.8) = \beta \approx 0.0000$$

and hence the power of the test is approximately 1. In Section 9 I shall discuss the import of this test in some detail.

4. NON-STANDARD TESTS: LIKELIHOOD TESTS

In contrast to the NP theory of testing is the non-standard theory of *likelihood tests.* As an example of such a theory I consider the one set out by Ian Hacking in [11]. The first notion to be defined is the *likelihood function* of a hypothesis h given observed data d. This is equal to the probability (or in the continuous case, the density) of d under the assumption that hypothesis h is true. That is

$$L(h/d) = P(d/h).$$

Likelihood functions differ from probability functions. The probability distribution of the result of observing random variable D with θ the fixed value of the relevant parameter, written $p(D/\theta)$, is a function over the space of possible results. As it defines a statistical distribution, summing over possible results (or integrating if D is a continuous variable), yields one. The likelihood function, in contrast, is a function of a statistical hypothesis (i.e., parameter) when the data is given. So now it is the data that is fixed as opposed to the parameter. This function does not define a statistical distribution and summing or integrating over possible hypotheses or parameters (even if these could be known) need not yield anything in particular – it clearly need not be one. The *likelihood ratio* of two hypotheses h_0 and h_1 given data d is defined as the ratio

$$L(h_1/d)/L(h_0/d)$$

and will be abbreviated as $r(d)$.

The basic intuition upon which likelihood inference is based is that the hypothesis with the greater support is the one having the greater likelihood on the data. This is formalized in the *law of likelihood* which states: Data d supports hypothesis h_0 better than hypothesis h_1 if the likelihood of h_0 given d is greater than that of h_1. In terms of relative likelihoods this becomes: Data d supports h_1 better than h_0 if $r(d) \geqslant 1$.

From the law of likelihood the theory of *likelihood tests* arises. According to Hacking

An hypothesis should be rejected if and only if there is some rival hypothesis much better supported than it is [11, p. 89].

Accordingly Hacking states [11, p. 90] that a simple hypothesis h_0 is *rejected at level c* in light of data d just in case there is a simple hypothesis h_0, consistent with the data, such that $r(d) \geqslant c$. For complex hypothesis H, d serves to reject H if for all simple hypotheses h_i in H, there is a hypothesis h_j that is consistent with the data such that $r(d) \geqslant c$.

The value of c in the likelihood test is the *critical ratio* (c-ratio) of the test and larger c values are taken to indicate more stringent tests. Associated with each likelihood test of hypothesis h_0 is a *rejection class* defined as [11, p. 90] the set of possible results for which there exists an alternative h_1 such that $r(d) \geqslant c$. The question as to what counts as "much better (or much less) supported" becomes a question as to what the appropriate c-ratio of a test is. According to Hacking, this is a matter of the quantity of evidence that is desired before rejecting a hypothesis. However, all c-ratios are not alike. Different tests having the same c-ratio can contain radically different quantities of evidence, and higher c-ratios need not mean more evidence. As a result, simply reporting that a hypothesis has been rejected at level c fails to provide sufficient information for comprehending an inference. Consider the following example.

EXAMPLE 4–1: Suppose 330 of the 1000 patients in hospital$_0$ and 990 of the 1000 patients in hospital$_1$ have disease X. A patient is randomly selected from three hospitals including these two, and it is observed whether or not he is suffering from X. The experiment may be described as the observation of a random variable D which takes on value d_1 if X is present, d_2 if it is not. Let θ be the probability of disease X. h_0 hypothesizes that the patient is from hospital$_0$ (i.e., $\theta = 0.330$) h_1, that he is from hospital$_1$ (i.e., $\theta = 0.990$).

The following chart summarizes the likelihoods for this experiment:

	d_1	d_2
$L(h_1/d_i)$	0.990	0.010
$L(h_0/d_i)$	0.330	0.670

Since $r(d_1) = 3$ a likelihood test rejects h_0 and accepts h_1 at this level. Having observed d_1 the likelihood statistician tells us to consider h_1 3 times better supported than h_0. However, such a statement is of questionable relevance for interpreting the result, as we would be told to favor h_1 at this level 33% of the time when in fact h_1 was false. This information shows up in the standard statistician's report that the size of the test is 0.33. In this way the rejection is distinguishable from other rejections at this level but where the size is much smaller. The likelihoodist, on the other hand, has no way of distinguishing this rejection from any other one at the same level. For him, any consideration of the probability of making certain errors is wholly irrelevant.

4.1. *Relevant Information*

Different attitudes towards inference give rise to different ideas as to what counts as relevant information. If information is processed solely by way of relative likelihoods, then what leaves them intact cannot contain information relevant for inference. This is embodied in the *likelihood principle* (LP), a notion about which statisticians admit confusion. In one of its weakest forms, the likelihood principle asserts that if two pieces of data yield proportional likelihoods for some hypothesis h, then they should compel the same statistical inference concerning h. A stronger version of the likelihood principle instructs us to look *only* at the likelihood function of observed data.

Both the likelihood and the Bayesian schools of inference subscribe to versions of this principle. Briefly, for a Bayesian, the LP follows from the fact that given a prior distribution, outcomes having proportional likelihood functions result in the same posterior – and inference is by the way of such posterior probabilities. As Savage explains:

> The evidential meaning of any experimental outcome is fully characterized by the likelihood function determined by the observations, without further reference to the structure of the experiment [23, p. 17].

While the present discussion is not specifically directed toward Bayesian inference, our criticism of likelihood inference bears upon Bayesian inference schemes as well. The manner in which it does will be discussed in Section 6.

The LP has the effect of rendering irrelevant considerations as to how the data were generated; that is, *sampling procedures, stopping rules*, as well as considerations of reliability (size), precision (power) and other criteria of SM, for the following reason. Inference according to the LP is conditional on the *particular realization d*. Unlike the SM, *d* is *not* viewed as one of several possible realizations of a random variable *D* (at least not for the purposes of making inferences from *d*). What might have happened is felt to be irrelevant for reasoning on the basis of what did happen. The reason, according to one Bayesian, is that "considerations about samples that have *not* been observed, are simply not relevant to the problem of how we should reason from the one that has been observed" [13, p. 200].

So, on this view only the particular sample observed is relevant, not the entire sample space. Accordingly, the sampling distribution which assigns probabilities to elements of the sample space is deemed irrelevant for inference. But each of the standard criteria (e.g., size and power) for formulating and evaluating inferences is based on the sampling distribution. Hence for followers of the LP each of these criteria is irrelevant. Lindley, one of the foremost Bayesians encapsulates this in the following remark:

> It is methods that are not based on the likelihood function that are suspect. In particular, unbiased estimates, minimum variance properties, sampling distributions, significance levels, power, all depend on something more – something that is irrelevant in Bayesian inference – namely the sample space [15, p. 436].

While followers of the LP criticize SM for considering the method of data collection relevant for inference, a follower of SM likewise considers the inability to take the sampling or data-generating method into account a fundamental weakness of the LP schools. Such criticisms do little more than express the fundamental disagreement between the rival accounts. What is wanted is an argument concerning the relevance of the sample space that does not simply presuppose the correctness of the favored inference philosophy. I shall attempt to provide such an argument from which I conclude that the sample space is relevant, and that considerations about what could have happened are relevant for reasoning about what did happen. This is particularly true if one is interested in prediction. The argument will proceed by showing that ignoring the sample space prevents any check on the frequency of errors. This leads to inferences which are incorrect with high probability – in extreme cases with probability one!

5. SPECIFICATION AFTER DATA (SAD) VS. SPECIFICATION BEFORE DATA (SBD)

If only likelihoods or ratios of likelihoods after the experiment are relevant, then such things as when one decides to stop sampling and when one designates the hypotheses to be tested are going to be irrelevant. The reason is that the likelihoods are not affected. Hence, calculation of the E–R measure, which in the case of Hacking is $r(d)$, is unaffected. For example, suppose the experiment is one of tossing coins in order to test a hypothesis about the probability that the coin lands heads. Consider two different plans for the production of data: (1) It is decided in advance to keep tossing until 6 heads are obtained and it happens to take exactly 100 tosses. (2) It is decided in advance to toss the coin 100 times and calculate the number of heads, and say the result is 6 heads. The likelihood of getting 6 heads, given that θ is the constant probability of heads, is, in case (1): $\binom{99}{94}\,\theta^6\,(1-\theta)^{94}$ and in case (2): $\binom{100}{6}\,\theta^6\,(1-\theta)^{94}$. Since these two likelihoods are proportional, a follower of the LP deems both experiments evidentially equivalent.[5]

In contrast, for SM both of these are quite different experiments. In the first case, prior to performing the experiment what is uncertain is the number of trials it will take to produce 6 heads. In the second case it is known that 100 trials are to be performed and what is unknown is the number of heads which will result. Since each case involves a different set of possible outcomes, each has a different probability distribution.[6] It follows that each case gives rise to a different probability of wrong inferences. As SM differentiates experiments by their associated error probabilities, it follows that it will have to distinguish between these two cases.

As long as the rule for stopping the production of data (i.e., the stopping rule) is deemed irrelevant, it is perfectly acceptable to continue taking observations until a desired likelihood value is obtained. Suppose the rule S for stopping an experiment is the following:

S: Sample until obtaining a d such that $L(h_0/d) \leqslant 1/c$

for a given simple hypothesis h_0 and for a positive and arbitrarily large value for c. That is, we can make $L(h_0/d)$ as small as we like using this stopping rule. Such an experiment will always, (with probability one) yield $L(h_0/d) \leqslant 1/c$, and so h_0 is given less than or equal to $1/c$ degree of support. In this way, as little support as is desired (for any c value of a likelihood test) may be assigned to h_0 even when h_0 is actually true! S in effect says

Continue sampling until h_0 is rejectible at level c.

The problem with this example is that the experiment which was to determine the degree of support for h_0 has h_0's support value built in through the stopping rule. This is a kind of likelihood version of the Liar's Paradox, and there is nothing in the likelihood approach to exclude such cases.

Now S permits h_0 to be rejected at any level c by a likelihood test since it is required only that there be *some* alternative (simple) hypothesis h_1 which is c times as likely on the data (i.e., $r(d) \geqslant c$). To serve as this alternative one can always choose the hypothesis which has the greatest likelihood on the data. This alternative will be one which says in effect that precisely what was observed had to have been observed. In other words, it will state that the data d arose from a process that had to produce d.

For example, if data d is the result of a coin tossing experiment, alternative h_1 may be that the probability of heads was 1 just on those tosses that resulted in a head. Using rule S we have $L(h_0/d) \leqslant 1/c$. Then, since the likelihood of this alternative h_1 given the data d is equal to 1, $r(d)$ is greater than or equal to $1/(1/c)$ which is just c. Having found a rival hypothesis at least c times as likely on the data as h_0, the likelihood test instructs us to reject h_0 and accept this alternative. The likelihood approach permits this since the ratio of likelihoods does not register a difference between cases where the hypotheses to test are SAD or SBD. It is possible to prove that stopping rule S permits the eventual rejection of *any* simple hypothesis h at any desired c-ratio in this manner, even if it is true.

There are two basic points which this goes to show. First, even a high c-ratio fails to provide information about which of two rival hypotheses is true, since hypotheses may be rejected by likelihood tests of *any* stringency in favor of some false hypothesis SAD. Hence likelihood inference can offer no guarantees about the reliability of its inferences. The second point concerns

SM. By insisting upon SBD, SM is able to guarantee it will not be wrong too often. In SAD cases, the logic of SM breaks down and no such guarantees can be claimed for its inferences. The following example will illustrate this. Here, it is seen that by permitting hypotheses to be SAD the probability of a false rejection is greater than the fixed size – indeed, it is equal to one. While the type of population is unusual, it must be remembered that the standard guarantee of a certain error rate is to hold regardless of the underlying population. So, in particular it should hold for this population.[7]

EXAMPLE 5–1 Imagine a population of 6 objects labelled *a, b, c, d, e, f,* each of which may possess some or all of 7 logically independent properties. Then we can talk about the proportion θ_i of the population that have property *i* where *i* ranges from 1 to 7. Let $P_i a$ express the claim that '*a* has property *i*'. The true state Λ of this population is as follows:

$$\begin{aligned} \Lambda = \ & P_1 a \,\&\, P_2 a \,\&\, P_3 a & & P_2 d \,\&\, P_4 d \,\&\, P_5 d \,\&\, P_6 d \\ & P_1 b \,\&\, P_3 b \,\&\, P_4 b & & P_2 e \,\&\, P_4 e \,\&\, P_5 e \,\&\, P_7 e \\ & P_1 c \,\&\, P_5 c \,\&\, P_6 c \,\&\, P_7 c & & P_3 f \,\&\, P_6 f \,\&\, P_7 f \end{aligned}$$

Suppose Λ is unknown and on the basis of a random sample of size 2 an inference about some property *i* is to be made. What is interesting about this population is that for any of the 15 possible samples of size 2 at least one of the properties is shared by both members of the sample, while in fact the true value of $\theta_i = 1/2$ for all *i*.

Having randomly selected 2 members of this population a SAD procedure may be: select a property *i* shared by both members and test the claim $\theta_i = 1/2$. To see how permitting the hypothesis to be SAD in this manner conflicts with standard tests, imagine an entire series of such experiments are made. While in each case the population may differ, suppose they each have the peculiar property that Λ has (i.e., each pair shares a property while $\theta_i = 1/2$ for all *i*). In each case the property *i* whose population proportion (θ_i) will be tested is specified as one for which the sample proportion is 1. Then, in each case the data is employed for evaluating a hypothesis for which it constitutes the most *misleading* evidence! That is, the chosen property *i* is that for which the observed proportion (1) differs most from the population (true) proportion, θ_i (1/2).

In contrast, an analogous standard test of $\theta_i = 1/2$ is to select a random sample and observe D_i, the proportion possessing a property *i*, where *i* is SBD. The set of possible outcomes of this experiment is the set of possible observed proportions of a preset property *i* in the pair selected, namely {0, 1/2, 1}.

In the SAD experiment above the set of possible outcomes is the set of *properties* $i(i = 1, 2, \ldots, 7)$ which may have a frequency of 1 in the observed sample. That a different experiment is involved from the SBD case does not show up in a likelihood test which reports only that $\theta_i = 1/2$ is only $1/5$ as strongly supported as $\theta_i = 1$ (where i is the property selected SAD). The difference does show up, however, in a measure of the reliability (i.e., size) of the test, as this is sensitive to the difference in the sample space.

In each standard test of $\theta_i = 1/2$ (for each of the i SBD) a critical region is specified of the form: $\{d_i: D_i \geqslant 1/2 + \delta\}$. δ is specified so that D_i falls into this region with a preset probability α. One can then calculate the number of such experiments needed in order to ensure $\delta < 1/2$. Then, regardless of the distribution of D_i, the test will not wrongly reject $\theta_i = 1/2$ more than $(100)\alpha\%$ of the time. (This follows from the fact that D_i is within δ of θ_i with probability $1 - \alpha$.)

However if the property to test were permitted to be SAD in the manner above, then, since D_i always equals 1, it always falls into the rejection region of the standard test. But since in fact the claim $\theta_i = 1/2$ is true for all i, one is lead to falsely reject it not with probability α but with probability 1! Hence, when the hypotheses are SAD as in Example 5–1, the standard test can no longer uphold its guarantees of reliability. More generally, whenever the tested property depends on the statistical features of the sample, a series of experiments can be constructed where the probability of incorrect inferences differs from a preset probability. In the analogous standard experiments the preset error rate is guaranteed for all underlying universes, and hence also for these specially constructed cases. This establishes the inconsistency between SM and theories which permit SAD procedures.

In light of our results we must reject as untenable the following claim by Hacking which he puts forward as an 'obvious fact':

If one hypothesis is better supported than another, it would usually be, I believe, right to call it the more reasonable [11, p. 28].

Although intuitively being well-supported seems like a desirable property for inferences to have, Hacking's notion of support does not embody our intuitive conception. For him, 'better supported' is equivalent to "more likely", but the best supported hypotheses SAD are generally false, and hence unreasonable. Even when dealing with two hypotheses neither of which is precisely true, the best supported hypothesis may not be the more reasonable.

For instance, it may be that a great deal of information about the underlying mechanism relating two quantities X and Y renders the hypothesis that

Y is some simple function of X correct within some small margin of error. Still, some other absurdly complex hypothesis may be more likely on the data although it contradicts theoretical background knowledge and as such is unreasonable. By advocating the acceptance of the hypothesis with the greatest likelihood on the data (or in general with the greatest E–R measure) one is advocating the most *ad hoc* hypothesis – that is, the one that differs least from the data. This is tantamount to suggesting that one not work with hypotheses at all but only with data!

But then what we have seems more useful for describing the data than for explaining it or for deriving predictions from it. If scientific models are to be used for predictions and explanations of phenomena as opposed to mere descriptions, then more than the specific data obtained must be taken into consideration. As a means of incorporating background information some likelihood theorists such as Edwards [4] make use of prior supports. The *Bayesians* make use of prior probabilities.

6. REMARKS ON BAYESIAN INFERENCE

It may be claimed that a follower of the Bayesian school of inference is not open to the kind of problems that SAD procedures pose for likelihoodists. Although both of these non-standard schools agree that likelihood functions exhaust the information from the experiment, the Bayesian also makes inferences dependent on something else; namely the *prior distribution* of the hypothesis. Without presenting the full theory of Bayesian testing, I shall now remark on how the problem of SAD bears upon Bayesian inference.

Bayesian inference proceeds by contemplating a (complete) set of hypotheses $h_i, \ldots h_k$ and using the sample data to obtain the probability of h_i conditional on this data by way of *Bayes Theorem*. The theorem states:

$$P(h_i/d) = \frac{P(d/h_i)P(h_i)}{\sum_{j=0}^{k} p(d/h_j)p(h_j)} \quad (i = 0, \ldots k).$$

It is seen that in order to calculate $P(h_i/d)$ one is required to have the values of $p(h_i)$ for all i. These are called the *prior probabilities*. The $p(h_i/d)$ are called the *posterior probabilities* – the basic E–R measure for Bayesians. One may use this measure to simply reject h_i if $P(h_i/d)$ is too small or to calculate a more elaborate test incorporating loses.[8]

The major controversy between Bayesians and non-Bayesians is not over the validity of this mathematically sound theorem, but over the use of prior probabilities of hypotheses. SM considers that the only sorts of things which have probabilities are events, such as the outcome 'heads' or '$D \geqslant d$'. SM does not assign probabilities to things considered fixed such as (i) random variables once their value has been observed, and (ii) hypotheses deemed either true or false and not able to be seen as the result of a probability experiment. When hypotheses can be construed as the result of a probability experiment, the relevant parameter θ may be construed as a random variable in contrast to its usual construal as a *fixed* value. In such cases, the standard methodologist has no qualms in applying Bayes Theorem.

An example of such a case is provided by Example 4–1. Suppose we have the additional information that all of the 1000 patients in the third hospital, hospital$_2$ have X. Hence $P(d_1/h_2) = L(h_2/d_1) = 1$, letting h_2 assert that the patient is from hospital$_2$ (i.e., $\theta = 1$). Then since there is an equal chance that the patient is from each of the 3 hospitals, θ may be viewed as a random variable taking on values 0.33, 0.99 and 1 with equal probability. Hence the priors $P(h_0) = P(h_1) = P(h_2) = 1/3$. Having observed d_1 the posterior $P(h_1/d_1)$ can be calculated using Bayes Theorem to be 0.43.

Typically, however, the hypothesis cannot be seen as the result of a probability experiment. Then the question as to what the prior probabilities mean and how they are to be obtained arises. The (subjective) Bayesian[9] sees the priors as measures of *degrees-of-belief* in hypotheses (in contrast to frequencies of occurrence of events). They are obtained by introspection or by cleverly considering how one would act in a series of bets. Lacking information about any of the hypotheses results in each being assigned an equal (or uniform) prior probability. This is known as the *principle of indifference.* Suppose in Example 4–1 one was not given the information that the patient was randomly selected from one of the three hospitals. Then, by virtue of one's ignorance, the 3 hypotheses would receive equal prior probability and the posterior $p(h_1/d_1)$ would still be 0.43. However, simply reporting this posterior (as a Bayesian would do) does not distinguish between the case where some random mechanism rendered the 3 hypotheses equiprobable and the case where one's ignorance does.

With respect to the problem of SAD two sorts of issues arise:

(1) *Priors SAD.* The Bayesian may argue that the sort of hypotheses SAD that result in objectionable inferences would be assigned a sufficiently small prior as to offset the high likelihood of h given d. But certainly there is no guarantee that all or even any Bayesian will happen to have a sufficiently low degree of belief in a hypothesis that happens to yield an objectionable

inference. Its objectionableness stems from the underlying population which presumably is unknown. This seems particularly true if priors are SAD and hence assigned (perhaps for several hypotheses) without knowing which the researcher plans to select, or for that matter, when. So while the Bayesian might avoid a high posterior in what turns out to be a false hypothesis SAD, it will be a matter of luck, and the Bayesian theory does not provide any means for distinguishing the lucky cases from the unlucky ones. The reason is that only the posterior is reported and not the prior nor the means by which the hypotheses, the data or the prior were generated.

In addition, there is nothing in the Bayesian philosophy itself to prevent the priors themselves from being SAD. The only principle is *coherency*, that is, not contradicting the probability axioms. It may be that if a Bayesian experimenter had been asked before the experiment for his prior probability for the claim: "all members of a random sample have a property *i*", that it would be quite low. However, once the experiment reveals a sample in which each element has the property *i*, he is likely to set his prior higher than he would have before the experiment. Then, the resulting posterior would be higher than if the prior were selected BD. Hence, if experimenting on the sort of population in Example 5–1, and both the hypotheses and the priors are SAD, the Bayesian continually assigns high posterior probabilities to false hypotheses.

It may be argued that one's prior degree of belief must always be SBD, but there seems to be nothing incoherent in having them be SAD. Indeed, there appears to be at least some cases in which Bayesians positively require that priors be SAD, such as when a subsequent observation indicates the prior is incorrect. As Lindley states, "Thus, if a prior leads to an unacceptable posterior then I modify it to cohere with properties that seem desirable in the inference" [15, p. 436]. However, Savage [23, p. 165] considers that a Bayesian who sets his prior AD on the basis of the very data which suggested the hypothesis is "in psychological danger of using the data twice." But it is still not clear that there are any strictly Bayesian grounds for ruling out double counting of data. (In the Lindley case it seems to be permitted.) In contrast, SM rules out such double use of data on the grounds that the probability of incorrect inferences is affected, as we have seen. But for a Bayesian any such consideration of error rates is irrelevant; his only official requirement is that one's total set of beliefs be coherent.

(2) *Hypotheses SAD.* Even if it is required that priors be SAD the Bayesian still appears to be open to the difficulties we have discussed concerning hypotheses SAD. Let us return to Example 5–1. Suppose that prior to the

experiment a Bayesian gives his priors for each property i for the claim: k out of the six have property i for $k = 0, 1, \ldots 6$. Suppose he has no information on the matter or for some other reason holds an equal degree of belief in each of the seven hypotheses for each i. (This could also be done using a continuous uniform prior.) As before, suppose it is decided to investigate claims about a property i shared by both members of a random sample selected from the six. The data d, the observed proportion having i is always (by the way i is selected) equal to 1. Having selected a sample, the Bayesian calculates the posteriors for the hypotheses about θ_i (the proportion that have property i), given $d = 1$.

For each i we have, $P(\theta_i = \frac{1}{2}/d) = 0.08$ and summing over the relevant hypotheses, $P(\theta_i > \frac{1}{2}/d) = 0.88$. So in each experiment the degree of belief assigned to H_i: $\theta_i > 1/2$ is 0.88. By repeating this type of experiment it is possible to evaluate hypothesis G_i: $P(\theta_i > 1/2) = 0.88$. If a sufficient number of repetitions are made, one's degree of belief in G_i continually increases towards one. This leads one to believe that in 88% of the experiments $\theta_i > 1/2$. In fact, however in 0 and not in 88% of the experiments is $\theta_i > 1/2$ (since $\theta_i = 1/2$). Thus one would be wrong about H_i many more times than expected.

It should be pointed out that less bizarre populations lead to the same sort of problem when hypotheses are SAD in the manner of Example 5–1, but with less dramatic results. Also, the above example may be elaborated upon to include utilities and thus bring it more in line with the typical type of Bayesian test.

I conclude that Bayesians, just as likelihoodists, should add to their philosophy some means for checking the rate of error or reliability of their inferences. Once reliability is seen to be relevant not only does the manner of data and hypotheses generation become relevant, but the manner of specification of priors as well.[10] Then each of these would have to be reported as information relevant to comprehending the experiment. If I am right, then this would have at least two undesirable consequences for the Bayesian philosophy. First, the Bayesian theory would forfeit one of its most attractive features: that of being a neat, simple, global account. Second, it would mean denying that a single probabilistic E–R measure, namely, the posterior distribution, is adequate for communicating all the relevant evidence for inference. This is to reject the LP. More specifically, to admit that the behavior of the E–R measure over different experiments is relevant, is to admit that not just the particular result of the particular experiment is relevant. In short, the one thing that the Bayesians insist is irrelevant (see p. 184) is admitted to be relevant, namely the sample space. This would seem to bring them closer to SM than most Bayesians would like.[11] However, I do not see how the Bayesian theory as it is presently formulated can provide means for testing the correctness of its tests.

I conclude that any purely formal attempt to relate data and hypotheses is inadequate. Behind such attempts seems to be the supposition that any *objective* inference scheme should lead from the same data and hypothesis to the same conclusion for any experimenter. This is precisely how the likelihood principle follows and with it the irrelevance of the data generation and order of hypotheses specification. From the standpoint of the formal quantitative E–R measure, it *is* irrelevant whether a hypothesis is SAD or SBD. It is not irrelevant, however, from the point of view of wanting hypotheses and theories that are at least approximately true and predictively reliable (to say nothing of interesting and informative).[12]

It should be pointed out that it is not that SBD methods are always more reliable than SAD ones. By formulating a suitable population it can be shown that the opposite is the case for that population. The real problem with SAD procedures, as I see it, is that whether or not they yield valid results is dependent upon which of the unknown hypotheses is true about the phenomena of interest. Hence a particular application of an SAD method cannot be identified as one of the cases resulting in a higher or a lower number of correct inferences. In contrast, by requiring SBD, SM can ensure a certain rate of error will not be exceeded regardless of which hypothesis is true. To my mind, it is this ability to guarantee validity that lies at the heart of an objective inference philosophy. This contrasts with E–R approaches which hold that objectivity is embodied in the requirement that the same data and hypothesis lead to the same conclusion.

7. VALIDATING THE EXPERIMENTAL MODEL

Even if the relevance of reliability and precision is granted, SM may be criticized as not being able actually to provide these guarantees. The standard test starts out with the assumption that we are observing a random variable that has a particular distribution and it remains only to determine the value of one of its parameters. But how do we know it actually has this distribution? The guarantees of reliability and precision hold only if the test statistic really has the assumed distribution, at least approximately.

This problem of validating the experimental model is shared by standard as well as non-standard philosophies. The likelihood function contains the relevant information from the data given a correctly specified statistical model, it does not contain all the information concerning the validity of this assumption. A separate test of the correctness of the underlying model is needed, ideally carried out before using it in a test based upon it.

Hacking admits [11, p. 222] that his theory is unable to test the underlying model. It cannot test claims before the data is in, nor can it test single hypotheses without alternatives. For a Bayesian to test whether an underlying model holds he is required to have a complete set of alternative hypotheses about how the model could fail as well as the necessary priors. It seems too much to require such detailed information be available at a stage where one is simply interested in whether a model is good enough to base an experiment upon.

Although the NP tests also involve alternative hypotheses, SM (despite what some people seem to think) does not consist solely of NP tests. Standard tests (e.g., Fisherian Significance Tests) are available to assess the 'goodness-of-fit' of a single hypothesis with data. Even NP tests do not require the specification of an alternative in order to run the test, but only to evaluate its power. As such, non-standard theorists may first find it useful to run certain standard tests in order to obtain reasonable alternatives as well as to explore with what sort of claims seem worth experimenting further. In addition to affording SM a means for building as opposed to only evaluating models, these single-hypothesis tests provide a means for testing the underlying model.

However, this seems to present a difficulty for SM. In order to test whether data follow a certain distribution one appears to be involved in testing a hypothesis based on the very data to be used in the test. But such hypotheses SAD were shown to prevent the guarantees of reliability. So it appears that in order to guarantee reliability (by validating the underlying model) the SM is required to carry out a procedure which precludes any such guarantees! Then, NP testing, which requires everything to be SBD, appears to be open to the following criticism by Rosenkrantz:

> Neyman–Pearson theory implies that data provide no support for an hypothesis suggested by those same data and not thought of or specified in advance. But matters become awkward when the hypothesis in question specifies one of the many ways in which the underlying model of the experiment could fail. For then we find ourselves forced to accept the conclusions erroneously based on that experiment merely because it was not designed to test the pertinent assumption of the underlying model! [22, p. 319].

According to Rosenkrantz, then, if a standard methodologist has run a coin-tossing experiment where it was assumed the tosses were independent, the aim being to test some value for the probability of heads, and then the data makes him question whether the tosses were really independent after all, he cannot take this into account in assessing the conclusion from the experiment. I will argue that he is not at all forced to do this, and that he can test the claim "the tosses are independent" after looking at the data without being guilty of an illegitimate SAD. Nor are additional coin tosses needed.

The data from the experiment designed to test 'P (heads) $= p$' may be *remodelled* in order to serve as data for testing 'the tosses are independent'. In the first experiment the possible data may be the different possible proportions of heads in n tosses, and there are $n + 1$ of these. The random variable D in this experiment is the proportion of successes, the order of occurrence being irrelevant. In the second test for independence, the number of successes d would now be regarded as fixed and the different data sets would be the different orders in which this number of successes might have occurred. (This is of relevance even if it is already known which one actually occurred.) There are $\binom{n}{d}$ such data sets.

Then, the significance test one might carry out is one based on the number of runs of heads (i.e., the number of consecutive 'heads') called a *runs test*. What is relevant for our purposes is that the data used for the test of the probability of heads is distinct from the data used in the subsequent test of independence. Hence, no illegitimate double-use of data is required.

A follower of SM need not insist that data never be used to test a hypothesis suggested by the same data in any scientific problem. He insists only that if the hypothesis is SAD one not use a procedure appropriate only for hypotheses SBD – at least not if one wants the associated claims about the errors of the inference to be valid.

Having argued that SAD should be avoided in order to guarantee reliability and precision of inferences, I am not claiming that any test with good 'error characteristics' (e.g., low size and high power) is a good test. Nor do I think that reporting the result of a test ('accept' or 'reject') even together with a description of the experiment and the two error probabilities is sufficient for correctly interpreting statistical tests. SM is also in need of additional extra-statistical principles both to specify and interpret tests – something which it has been reluctant to provide, at least explicitly. In what follows I shall merely sketch the sort of supplements I have in mind. They are set out in more detail in [17].

8. THE RATIONALE OF TESTING

Much of the confusion and controversy concerning statistical methodologies stems from the fact that the formalism in which they are presented has little bearing on how they are actually used to make informative scientific inferences. Neyman–Pearson tests are presented as a kind of recipe for making 'optimal' inferences. One simply fixes the size α of a test, finds the most powerful test having a given size and then accepts or rejects. If asked why anyone should find it desirable to do this the rationale given by Neyman (see [21]) is this: If one 'behaves' in this way one will incorrectly reject h_0 not more than 100α percent of the time and incorrectly accept h_0 not more than 100β percent of the time, (for α, β the probabilities of type I and type II errors respectively).

However, Neyman's rationale may be criticized. Tests which are 'best' on the standard criteria need not really be best.[13] One need not look further than a simple coin-tossing experiment to find an example of such a test.

EXAMPLE 8–1: Test whether θ the probability of heads is $0.35(h_0)$ or $0.10(h_1)$ by tossing a coin 4 times, and observing D, the proportion of heads. Letting $d_i = i/4$, $i = 0, \ldots 4$, $d_0 - d_4$ are the five possible proportions. The size α is set at 0.18. The likelihoods are as follows:

	d_0	d_1	d_2	d_3	d_4
$L(h_0/d_i)$	0.17	0.38	0.31	0.11	0.01
$L(h_1/d_i)$	0.65	0.29	0.04	0.003	0.001

The most powerful test of size 0.18 is test T_1 which instructs one to reject h_0 iff d_0 or d_4 occurs. Yet if d_4 occurs (i.e., all heads) T_1 rejects $\theta = 0.35$ and accepts $\theta = 0.10$ although d_4 clearly makes the former more plausible.

A less powerful, but intuitively more satisfactory test is T_2 which rejects h_0 iff d_0 occurs. However, to select text T_2 upon seeing it is more sensible than T_1 is not strictly to follow NP test criteria. (While the size of T_2 is slightly less than for T_1, T_1's greater power renders T_1 the best test of its size.)

T_2 is a more sensible test as it is based on a more sensible test statistic, namely one which measures the discrepancy between the observed proportion d and the proportion hypothesized by h_0 (i.e., 0.35). A plausible basis for choosing this discrepancy measure is that of (Euclidean) *distance*. Letting $T(D)$ be $D - 0.35$ it is seen that only d_0 yields a value of $T(D)$ less than -0.1. Hence test T_2 is equivalent to the rule: reject h_0 iff $T(D) < -0.1$ for $T(D)$ as defined. This test statistic provides a plausible ranking of the observations in order of how discrepant they are from h_0 (in the direction of h_1).

What I am claiming is that while tests with good error characteristics may coincide with good (or even best) tests, the error values are not themselves the reason for the test's goodness. It is only when the error rates are associated with a reasonable test statistic – one that is some sort of a measure of discrepancy – that good tests result. It is not low error rates in the long run that justify tests, since as critics of SM point out, one can do well in the long run but terribly in a particular case. Yet the rationale for NP tests according to Neyman is long run error rates alone. I would argue that the value of error rates is that combined with sensible test statistics they help ensure that the test will reject hypotheses just in case the data indicates the existence of a discrepancy from the null. One does this by ensuring that an extreme value of a test statistic (one leading to a rejection) be associated with a discrepancy of desired size.

Hence, the importance of being able to guarantee the reliability and precision (size and power) of an inference independently of unknown parameter values and of priors is not directly a concern with how often one will be right. Rather, the importance stems from a desire to guarantee that *this* test is picking up all and only discrepancies of a certain size. Ideally the discrepancies being detected will be of non-negligible size. However, even if this is not the case, the standard test need not yield faulty inferences as long as it is possible to ascertain after the experiment what sort of discrepancies were being detected. This is possible by guaranteeing that preset levels of reliability and precision are met. We have seen how SAD procedures permit actual error rates to differ greatly from preset ones. This has the effect of permitting the test to detect discrepancies of a size different from the intended one. Hence, from the present point of view SAD is objectionable in that it prevents one from objectively determining the type of discrepancy a test detects. In contrast, SM does permit such an objective determination and it would be desirable to report in addition to the reliability and precision, the type of discrepancies detected.

Standard tests are often criticized because with a large enough sample one can ensure reaching any designated size and power in rejecting a null hypothesis. This is because as sample size increases the standard deviation of the statistic observed decreases, being inversely proportional to the sample size. Hence, with a large enough sample, even a highly statistically significant difference may reflect a tiny discrepancy from the null. This leads to faulty inferences because typically statistical inferences are not distinguished from the subsequent scientific inferences based upon them. Upon distinguishing these two inferences one does not automatically interpret a rejection of a statistical hypothesis as the rejection of a substantive scientific hypothesis.

Tests should be supplemented with rules for interpreting the scientific or substantive significance of statistically significant results. What I have in mind may be illustrated by referring back to Example 3–1.

This test is a 'best' test for the given size 0.03 and it has extremely high power, practically one. Still, the question arises as to what information this test is providing. The following graphs indicate the relationship between the size and power of the tests:

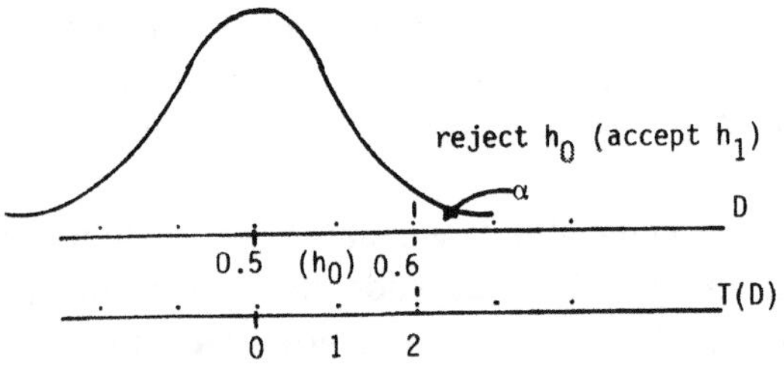

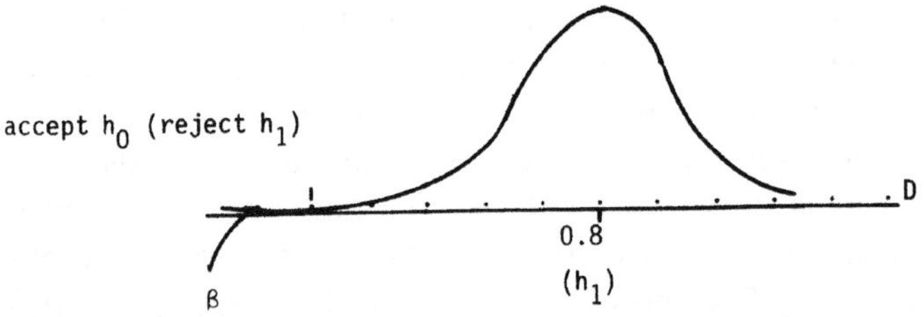

Since β is practically zero and is many times smaller than α, the test is sensitive to relatively small departures (in the direction of h_1) from h_0, and insensitive to large departures from h_1 (in the direction of h_0). In order to correctly test what information the test is providing one may calculate what I shall call the *effective discrepancy*. This is the size of the discrepancy that the test, in effect, is detecting. To obtain it one calculates for what alternative hypothesis the test would have the same probability of making a type I as it would a type II error. In this case it is approximately $\theta = 0.69$. This means that observations which typically arise given that $\theta = 0.69$ are taken as rejecting $\theta = 0.5$ and accepting $\theta = 0.8$ which may or may not be a desirable thing to do. The point is, instead of merely reporting the null was rejected, by a test with such and such size and power, the effective discrepancy should be reported. (A report of the sample size provides enough information together with the size and power to calculate the effective discrepancy but it is not itself as illuminating.)

Another way of correctly assessing the results of a test is to calculate the appropriate number of observations for which the test in question would be a fair one between the two given hypotheses. By a fair test I mean one where $\alpha = \beta$. In this case, had the sample size been only 35, the test would have been a fair test between $\theta = 0.5$ and $\theta = 0.8$. Hence, rejecting a test with this size and power given it had only 35 observations is actually more significant than rejecting it at this level with the 100 observations. This may seem counter-intuitive. Indeed, researchers typically believe that a rejection based on a larger sample size is more significant than one based on a smaller one. The reason stems from the fact that more instances of something is taken to provide more evidence. SM is sometimes criticized for seeming to go against this scientific principle, but in fact it does not do so. More instances do provide better evidence, but in adding to the sample size one is not adding instances but rather changing what counts as a *single* instance (e.g., the proportion of heads out of 35 as opposed to out of 100 tosses).

Having found the appropriate sample size to be smaller than the size of the sample actually made, should one simply throw out the extra data that is already available? No, the purpose of calculating the appropriate sample size even after the experiment is in order to comprehend the significance of the results obtained. By noting that the appropriate sample size is much smaller than the actual one it is seen that the test is detecting discrepancies which may be of negligible substantive importance. It is not part of the statistical test to decide what counts as a non-negligible discrepancy, but I do think it should provide the means for ascertaining the sort of discrepancy detected for two reasons: (1) for the sake of designing tests that detect discrepancies one is interested in and (2) for correctly evaluating the substantive significance of a test already made.

In other words, determining the effective discrepancy or the appropriate sample size provides a means for testing the appropriatness of statistical tests. In interpreting an inference one can argue that through inappropriate test specifications the test is rendered either too sensitive or too insensitive. However, even inferences based on tests with inappropriate sensitivity need not be illegitimate. If, for example, the test in Example 3–1 yields an inference not to reject h_0, the inference would be considered sound. The reason is that no discrepancy was detected by a test that had an abundance of power to detect relatively minor discrepancies. In general, if the sample size is so large as to discriminate an alternative closer to the test hypothesis than one that is scientifically interesting, and yet the null hypothesis is not rejected, the inference is sound (barring any other inadmissible moves). That is,

> If a test with an overabundance of power to detect a discrepancy fails to do so, then the conclusion is sound.

Similarly, if a sample size that is small relative to the appropriate sample size is used, and the test hypothesis is rejected in favor of the alternative, then the inference appears sound. That is,

> If a test which could not reasonably be expected to detect a discrepancy does detect one, then its conclusion is sound.

Under any other circumstances, however, inferences based on tests with incorrect sensitivities provide reason to question the relevance of the result. If an inference is based on a test with too many observations (relative to a balance between α, β and the alternative hypothesis) and the null is rejected, the reason for the rejection may be a trivial perturbation of the experiment. It may be the result of no real difference at all, or of one that is scientifically insignificant.

By viewing tests as tools for detecting discrepancies one is also lead to design tests with a view to having them detect all and only discrepancies of a certain size. It is up to the scientist to specify the kind of discrepancies about which he considers worth knowing. It is up to the test to detect all and only the sort of discrepancies it is commanded to detect. The value of the standard tests, I have argued, is that if used correctly they may be seen to function in this way. In short, they may be seen to provide objective standards for measuring discrepancies.

9. CONCLUSION

Our conclusion about what counts as a good test differs markedly from Hacking who claims

> A good test for any hypothesis, statistical or otherwise, is one which rejects it only if another hypothesis is much better supported [11, p. 109].

We have seen that one can always find a hypothesis better supported in this sense (i.e., more likely) by selecting the most *ad hoc* hypothesis SAD. Instead, a good test for a statistical hypothesis is one which rarely fails to find a discrepancy of a given size (i.e., rarely accepts a hypothesis) unless no non-trivial discrepancy exists, and rarely finds a discrepancy of a given size (i.e., rarely rejects a hypothesis) unless a non-trivial discrepancy exists.

The reason for banning illegitimate SAD procedures is to guarantee claims about reliability and precision – but these are not ends in themselves, at least not in a scientific context. Their importance stems from a desire to check that a test is doing what it is supposed to be doing – that is, detecting discrepancies of a certain size. It provides a way of testing the validity of the test. Without such a means for testing statistical testing tests fail to be objective and to link up with truth.

Virginia Polytechnic Institute and State University

NOTES

* I am very grateful to I. J. Good and Joseph Pitt for useful comments on earlier versions of this paper, and to George Shapiro for mathematical advice.

1 The most complete inference theory in this category is the one proposed by Carnap [3] and extended by his followers, most notably, Hintikka.

2 I shall here omit consideration of cases where more than one unknown parameter exists.

3 The standard deviation of D, σ_D is the square root of the average deviation (or *variance*) of D from its mean θ. The variance of D, σ_D^2 is

$$\sum_i (d_i - \theta)^2 P(D = d_i)$$

for D a discrete random variable with values d_i. (In the continuous case the summation is replaced by an integral.)

4 This experiment may be modeled binomially provided that each toss (i) can yield only one of two results (heads, tails), (ii) is independent, (iii) has the same probability θ of landing heads.

The probability of getting k heads out of n tosses (i.e., $D = k/n$) is

$$\binom{n}{k} \theta^k (1 - \theta)^{n-k}$$

The standard deviation of D,

$$\sigma_D = \sqrt{\theta(1 - \theta)} / n$$

5 That is, the likelihoodist considers the decision as to when to stop sampling irrelevant. This is referred to as the *irrelevance of optional stopping.*

6 In case (1) the distribution is the *negative binominal* in case (2) it is the *binomial.*

7 Populations of this sort are discussed more generally in [6] to make a similar point but concerning Bayesian estimation.

8 In testing h_0: $\theta = \theta_0$ vs h_1: $\theta = \theta_1$ where the priors $p(h_0)$, $p(h_1)$ are given and a, b are the losses associated with a mistaken rejection and a mistaken acceptance of h_0 respectively, the Bayes test is to reject h_0 iff

$$\frac{ap(d/h_0)p(h_0)}{bp(d/h_1)p(h_1)} < 1$$

given data d. Given any Neyman–Pearson test one can find losses and priors that make it equivalent to a Bayes test.

[9] While there are those who call themselves objective, or empirical Bayesians, I will here understand Bayesian to mean subjective Bayesian.

[10] In reporting the manner in which priors were specified it is not only relevant to report whether they were SAD or SBD. It is also relevant to the reliability of the result to know whether the priors have resulted from an experiment or from a prior degree of belief without an experiment. The former case appears more reliable.

[11] Possible exceptions would be those willing to hold a Bayesian philosophy which compromises between standard and non-standard views, most notably I. J. Good.

[12] Bayesians often admit that their philosophy is not about truth but simply provides a theory for personal decision making. As Lindley puts it, "The Bayesian Theory is about *coherence*, not about right or wrong" [16, p. 359].

[13] In criticizing Neyman's rationale Hacking [11, p. 96] gives an example of a test which is 'better' than some other test on NP criteria but which is clearly not better once the data is in. However this test is not the 'best' NP test and hence would not be recommended by the NP theory. As such his criticism does not present a problem for the NP theory. The example I present (Example 8–1) does.

REFERENCES

[1] Barnard, G. A., 'The logic of statistical inference', *British Journal of the Philosophy of Science* **23** (1972), 123–190.

[2] Carnap, R., *Logical Foundations of Probability*, 2nd edition, University of Chicago Press, Chicago, 1962.

[3] Cunnyngham, J., 'Econometric model construction and predictive testing', in *Problems and Issues in Current Econometric Practice*. Ed. K. Brunner, The Ohio State University, Ohio, 1972, pp. 238–261.

[4] Edwards, A. W. F., *Likelihood*, Cambridge University Press, Cambridge, 1972.

[5] Fisher, R. A., *Statistical Methods and Scientific Inference*, 2nd edition, Oliver and Boyd, Edinburgh, 1959.

[6] Giere, R. N., 'Bayesian statistics and biased procedures', *Synthese* **20** (1969), 371–387.

[7] Godambe, V. P., and Sprott, D. A. (eds.), *Foundations of Statistical Inference*, Holt, Rinehart and Winston of Canada, Toronto, 1971.

[8] Good, I. J., 'Probability or Support?', *Nature* **213** (1967), No. 5073, 233–234.

[9] Good, I. J., 'The Bayesian influence, or how to sweep subjectivism under the carpet', in [12], pp. 125–174.

[10] Goodman, N., *Fact, Fiction, and Forecast*, The Bobbs Merrill Company, Inc., New York, 1965.

[11] Hacking, I., *Logic of Statistical Inference*, Cambridge University Press, Cambridge, 1965.

[12] Harper, W. L., and Hooker, C. A. (eds.), *Foundations of Probability Theory, Statistical Inference, and Statistical Theories of Science*, Vol. II, D. Reidel Publishing Co., Holland, 1976.

[13] Jaynes, E. T., 'Confidence intervals vs Bayesian intervals', in [12], pp. 175–213.

[14] Lieberman, Bernhardt (ed.), *Contemporary Problems in Statistics*, Oxford University Press, New York, 1971.
[15] Lindley, D. V., 'The estimation of many parameters', in [7], pp. 435–447.
[16] Lindley, D. V., 'Bayesian statistics', in [12], pp. 353–362.
[17] Mayo, D., *Philosophy of Statistics*, Doctoral dissertation, University of Pennsylvania, 1979.
[18] Mood, A. M., Graybill, F. A., and Boes, D. C., *Introduction to the Theory of Statistics*, 3rd edition, McGraw-Hill, Inc., New York, 1963.
[19] Morrison, D. E. and Henkel, R. E. (eds.), *The Significance Test Controversy – A Reader*, Aldine Publishing Company, Chicago, 1970.
[20] Neyman, J., *First Course in Probability and Statistics*, Henry Holt, New York, 1950.
[21] Neyman, J. and Pearson, E. S., 'On the problem of the most efficient tests of statistical hypotheses', *Philosophical Transactions of the Royal Society* **A231** (1933), 289–337.
[22] Rosenkrantz, R., 'The significance test controversy', *Synthese* **26** (1973), 304–321.
[23] Savage, L. J., *The Foundations of Statistics*, Wiley and Sons, Inc., New York, 1954.

INDEX

THE UNIVERSITY OF WESTERN ONTARIO SERIES IN PHILOSOPHY OF SCIENCE

A Series of Books in Philosophy of Science, Methodology, Epistemology, Logic, History of Science, and Related Fields

1. J. Leach, R. Butts, and G. Pearce (eds.), *Science, Decision and Value*. Proceedings of the Fifth University of Western Ontario Philosophy Colloquium, 1969. 1973, vii + 213 pp.
2. C. A. Hooker (ed.), *Contemporary Research in the Foundations and Philosophy of Quantum Theory*. Proceedings of a Conference held at the University of Western Ontario, London, Canada, 1973. xx + 385 pp.
3. J. Bub, *The Interpretation of Quantum Mechanics*. 1974, ix + 155 pp.
4. D. Hockney, W. Harper, and B. Freed (eds.), *Contemporary Research in Philosophical Logic and Linguistic Semantics*. Proceedings of a Conference held at the University of Western Ontario, London, Canada. 1975, vii + 332 pp.
5. C. A. Hooker (ed.), *The Logico-Algebraic Approach to Quantum Mechanics*. 1975, xv + 607 pp.
6. W. L. Harper and C. A. Hooker (eds.), *Foundations of Probability Theory, Statistical Inference, and Statistical Theories of Science*, 3 Volumes. Vol. I: *Foundations and Philosophy of Epistemic Applications of Probability Theory*. 1976, xi + 308 pp. Vol. II: *Foundations and Philosophy of Statistical Inference*. 1976, xi + 455 pp. Vol. III: *Foundations and Philosophy of Statistical Theories in the Physical Sciences*. 1976, xii + 241 pp.

8. J. M. Nicholas (ed.), *Images, Perception, and Knowledge.* Papers deriving from and related to the Philosophy of Science Workshop at Ontario, Canada, May 1974. 1977, ix+309 pp.
9. R. E. Butts and J. Hintikka (eds.), *Logic, Foundations of Mathematics, and Computability Theory.* Part One of the Proceedings of the Fifth International Congress of Logic, Methodology and Philosophy of Science, London, Ontario, Canada, 1975. 1977, x+406 pp.
10. R. E. Butts and J. Hintikka (eds.), *Foundational Problems in the Special Sciences.* Part Two of the Proceedings of the Fifth International Congress of Logic, Methodology and Philosophy of Science, London, Ontario, Canada, 1975. 1977, x+427 pp.
11. R. E. Butts and J. Hintikka (eds.), *Basic Problems in Methodology and Linguistics.* Part Three of the Proceedings of the Fifth International Congress of Logic, Methodology and Philosophy of Science, London, Ontario, Canada, 1975. 1977, x+321 pp.
12. R. E. Butts and J. Hintikka (eds.), *Historical and Philosophical Dimensions of Logic, Methodology and Philosophy of Science.* Part Four of the Proceedings of the Fifth International Congress of Logic, Methodology and Philosophy of Science, London, Ontario, Canada, 1975. 1977, x+336 pp.
13. C. A. Hooker (ed.), *Foundations and Applications of Decision Theory*, 2 volumes. Vol. I: *Theoretical Foundations.* 1978, xxiii+442 pp. Vol. II: *Epistemic and Social Applications.* 1978, xxiii+206 pp.
14. R. E. Butts and J. C. Pitt (eds.), *New Perspectives on Galileo.* Papers deriving from and related to a workshop on Galileo held at Virginia Polytechnic Institute and State University, 1975. 1978, xvi + 262 pp.
15. W. L. Harper, R. Stalnaker, and G. Pearce (eds.), *Ifs. Conditionals, Belief, Decision, Chance, and Time.* 1980, ix + 345 pp.
16. J. C. Pitt (ed.), *Philosophy in Economics.* Papers deriving from and related to a workshop on Testability and Explanation in Economics held at Virginia Poly-Technic Institute and State University, 1979. 1981.

GPSR Compliance

The European Union's (EU) General Product Safety Regulation (GPSR) is a set of rules that requires consumer products to be safe and our obligations to ensure this.

If you have any concerns about our products, you can contact us on ProductSafety@springernature.com

In case Publisher is established outside the EU, the EU authorized representative is:

Springer Nature Customer Service Center GmbH
Europaplatz 3
69115 Heidelberg, Germany

Batch number: 10823775

Printed by Printforce, the Netherlands